Strategic Environmental
Assessment for Policies

E N V I R O N M E N T
A N D
D E V E L O P M E N T

A fundamental element of sustainable development is environmental sustainability. Hence, this series was created in 2007 to cover current and emerging issues in order to promote debate and broaden the understanding of environmental challenges as integral to achieving equitable and sustained economic growth. The series will draw on analysis and practical experience from across the World Bank and from client countries. The manuscripts chosen for publication will be central to the implementation of the World Bank's Environment Strategy, and relevant to the development community, policy makers, and academia. Topics addressed in this series will include environmental health, natural resources management, strategic environmental assessment, policy instruments, and environmental institutions.

Also in this series:
International Trade and Climate Change: Economic, Legal, and Institutional
 Perspectives
Poverty and the Environment: Understanding Linkages at the Household Level

Strategic Environmental Assessment for Policies

An Instrument for Good Governance

Kulsum Ahmed and
Ernesto Sánchez-Triana, Editors

THE WORLD BANK
Washington, DC

© 2008 The International Bank for Reconstruction and
Development/The World Bank
1818 H Street, NW
Washington, DC 20433
Telephone 202-473-1000
Internet www.worldbank.org
E-mail feedback@worldbank.org

All rights reserved.

1 2 3 4 :: 11 10 09 08

This volume is a product of the staff of the International Bank for
Reconstruction and Development / The World Bank. The findings, interpretations, and conclusions expressed in this volume do not necessarily
reflect the views of the Executive Directors of The World Bank or the
governments they represent.

The World Bank does not guarantee the accuracy of the data included
in this work. The boundaries, colors, denominations, and other information shown on any map in this work do not imply any judgement on the
part of The World Bank concerning the legal status of any territory or the
endorsement or acceptance of such boundaries.

R I G H T S   A N D   P E R M I S S. I O N S

The material in this publication is copyrighted. Copying and/or transmit-
ting portions or all of this work without permission may be a violation of
applicable law. The International Bank for Reconstruction and
Development / The World Bank encourages dissemination of its work and
will normally grant permission to reproduce portions of the work promptly.

For permission to photocopy or reprint any part of this work, please
send a request with complete information to the Copyright Clearance Center
Inc., 222 Rosewood Drive, Danvers, MA 01923, USA; telephone: 978-750-
8400; fax: 978-750-4470; Internet: www.copyright.com.

All other queries on rights and licenses, including subsidiary rights,
should be addressed to the Office of the Publisher, The World Bank,
1818 H Street NW, Washington, DC 20433, USA; fax: 202-522-2422;
e-mail: pubrights@worldbank.org.

ISBN-13: 978-0-8213-6762-9
eISBN-13: 978-0-8213-67636
DOI: 10.1596/978-0-8213-6762-9

Library of Congress Cataloging-in-Publication Data
Strategic environmental assessment for policies: an instrument for good
governance / Kulsum Ahmed and Ernesto Sánchez-Triana, editors.
       p. cm.
   "This edited book derives from the background papers originally prepared
as the basis for a World Bank study published in 2005, titled Integrating
Environmental Considerations in Policy Formulation: Lessons from Policy-
Based SEA Experience (Report No. 32783)."
   ISBN 978-0-8213-6762-9 — ISBN 978-0-8213-6763-6 (electronic)
   1. Environmental policy. 2. Environmental impact analysis. 3. Sustainable
development. 4. Strategic planning—Environmental aspects. I. Ahmed,
Kulsum, 1964- II. Sánchez Triana, Ernesto.
   GE170.S766 2008
   333.72—dc22
                                                      2008001297

Cover photo/illustrations:
Waterfall in Guyana/Louie Psihoyos/Corbis; Girl in Central African
Republic/Giacomo Pirozzi/Panos

Cover design:
Auras Design, Silver Spring, Maryland

CONTENTS

About the Editors and Authors

Kulsum Ahmed is lead environmental specialist at the World Bank and team leader for the Environmental Institutions and Governance program (which includes, among other activities, the SEA program) and Environmental-Health program in the Environment Department. She has considerable experience as an operations task manager and led the team that prepared the Bank's first structural adjustment loan to integrate environmental considerations in key sectors of a country's economy. She is the World Bank's representative on the OECD Development Assistance Committee's task team on strategic environmental assessment. Ms. Ahmed has authored numerous publications on energy, the environment, health, strategic assessment, and industrial pollution. She holds a Ph.D. from Imperial College, London, and a natural sciences degree from Cambridge University.

Ernesto Sánchez-Triana is senior environmental engineer at the World Bank South Asia Region. Before joining the Bank he taught at Colombia's National University, managed the Special Division of Environmental Policy in Colombia's Department of National Planning, and worked at the Inter-American Development Bank. At the World Bank, he has led the preparation of policy-based loans that aim to incorporate environmental considerations into economic and sectoral policies, as well as the conduction of comprehensive assessments to identify and correct the institutional weaknesses that contribute to environmental degradation. Mr. Sánchez-Triana is the author of numerous publications on environmental and energy policy, political economy, and the use of economic instruments for environmental protection. He holds a Ph.D from Stanford University.

Harry Blair is a senior research scholar and lecturer in political science at Yale University.

Alnoor Ebrahim is a visiting associate professor at the John F. Kennedy School of Government and the Wyss visiting scholar at the Harvard Business School.

Martha S. Feldman holds the Johnson Chair for Civic Governance and Public Management at the University of California–Irvine.

Caroline Kende-Robb is a sector manager in the Social Development Department at the World Bank.

Anne M. Khademian is an associate professor at the Center for Public Administration and Policy at the School for Public and International Affairs, Virginia Polytechnic Institute and State University.

Richard D. Morgenstern is a senior fellow at Resources for the Future.

Leonard Ortolano is the UPS Foundation professor in the department of civil and environmental engineering at Stanford University.

Warren A. Van Wicklin III is an international development consultant.

Acknowledgments

This edited book derives from the background papers originally prepared as the basis for a World Bank study published in 2005, titled *Integrating Environmental Considerations in Policy Formulation: Lessons from Policy-Based SEA Experience* (Report No. 32783). In converting these papers to stand-alone chapters in this book, many people have played an important role. These include the chapter authors, who worked diligently to enrich their original manuscripts with additional case studies drawn, in particular, from developing countries; the formal peer reviewers, who provided valuable comments to improve this manuscript; and the many other colleagues who provided important resources, insights, and research assistance in the preparation of these chapters.

Formal chapter peer reviewers included Peter Croal (Canadian International Development Agency); Linda DeLeon (University of Colorado); Margaret Keck (Johns Hopkins University); Bjorn Larsen (economist and independent consultant), Leonard Ortolano (Stanford University); Nancy Roberts (Naval Postgraduate School); Rob Verheem (Netherlands EIA Commission); and Sameer Akbar, Yewande Awe, Giovanna Dore, Fernando Loayza, Giovanni Ruta, Jeff Thindwa, and Shahid Yusuf (all of the World Bank). This work was carried out under the overall guidance of Laura Tlaiye, Sector Manager, and James Warren Evans, Director, both of the Environment Department at the World Bank.

Finally, the editors would also like to thank Jim Cantrell, Mary Fisk, and Patricia Katayama for their support during the publication and dissemination phase of this book.

This volume is dedicated to Professor Dennis Anderson, whose talent for economics and humble approach to life combine to inspire all development professionals who have had the good fortune of knowing and interacting with him. His standard of professionalism and humanity is one to which we all aspire.

Abbreviations

ANC	African National Congress
APA	*área de proteção ambiental*, or environmentally protected area
BAPEDAL	*Badan Pengendalian Dampak Lingkungan*, or Environmental Impact Management Agency
BOD	biochemical oxygen demand
BPP	*Bangun Praja* Program, or Good Environmental Governance Program
CARs	*corporaciones autónomas regionales*, or regional development corporations
CEA	Country Environmental Analysis
CEPD	Centre for Educational Policy Development
CNG	compressed natural gas
COD	chemical oxygen demand
CONNEPP	Consultative National Environmental Policy Process
CRA	comparative risk analysis
CSE	Centre for Science and the Environment
CSIR	Council of Scientific and Industrial Research
CSO	civil society organization
DALY	disability-adjusted life year
DFAIT	Department of Foreign Affairs and International Trade
DPRD	*Dewan Perwakilan Rakyat Daerah*, or local elected council
EIA	environmental impact assessment
ENVIRONET	Committee Network on Environment and Development Cooperation
EPA	Environmental Protection Agency
EPCA	Environment Pollution Prevention and Control Authority
E-test	environmental test
GDP	gross domestic product
ITAC	International Trade Advisory Committee
KLH	*Kementerian Negara Lingkungan Hidup*, or Ministry of the Environment
MPRP	Mongolian People's Revolutionary Party
NAFTA	North American Free Trade Agreement
NGO	nongovernmental organization
NTB	National Training Board
OECD	Organisation for Economic Co-operation and Development
PAT	*plan de acción trienal*, or three-year action plan
PLSA	Participatory Living Standards Assessment
PM	particulate matter

PROKASIH	Clean River Program
PROPER	Program for Pollution Control, Evaluation, and Rating
PSIA	Poverty and Social Impact Analysis
SAGIT	Sectoral Advisory Group on International Trade
SEA	Strategic Environmental Assessment
SOP	Sectoral Operational Program for Tourism and the Spa Industry
TOR	terms of reference
UNECE	United Nations Economic Commission for Europe
USAID	U.S. Agency for International Development
WALHI	*Wahana Lingkungan Hidup*, or Friends of the Earth
WHO	World Health Organization

C H A P T E R   1

SEA and Policy Formulation

Kulsum Ahmed and Ernesto Sánchez-Triana

SOUND ENVIRONMENTAL MANAGEMENT is particularly important for the poor, whose lives can be transformed by development. Good management of the environment and natural resources protects health, reduces vulnerability to natural disasters, improves livelihoods and productivity, spurs economic growth based on natural resources, and enhances human well-being. Environmental risk factors play a role in more than 80 diseases, and injuries and account for more than a third of disease in children under five. Better environmental management could reportedly prevent more than 94 percent of deaths from diarrheal diseases and 40 percent of deaths from malaria, saving the lives of as many as 4 million children a year (Pruss-Ustun and Corvalan 2006).

Between 1980 and 2000 more than 1.5 million people died in floods, volcanic eruptions, tropical storms, landslides, droughts, and other natural disasters (UNDP 2004). The tsunami in the Indian Ocean in December 2004 killed more than 200,000 people, left more than 1.5 million homeless, and destroyed more than US$6 billion of physical capital (World Bank 2006). Deaths and economic losses caused by natural disasters are far greater in poor countries than in rich ones.

Kulsum Ahmed is lead environmental specialist and team leader of the Environmental Institutions and Governance program at the World Bank. Ernesto Sánchez-Triana is senior environmental engineer at the World Bank.

High vulnerability to natural disasters in poor countries is linked with the lack of warning and early response systems and inadequate integration of environmental considerations in regional and urban development.

In the poorest countries of Africa, Asia, and Latin America, forest, agricultural, and biological resources contribute significantly to the economy of poor rural households.[1] Some 200 million indigenous people depend on forests for their livelihood, food, medicine, and shelter (IUCN 2006). The environment constitutes a fundamental driver of employment and economic growth. Tourism in Costa Rica (mainly ecotourism), for example, accounts for 8.4 percent of GDP, generating about 72 percent of national monetary reserves and supporting 140,000 jobs (UNEP 2007). In Mexico, where tourism accounts for 9 percent of GDP, environmental quality is a key determinant of tourists' choice of destination (SECTUR 2002; World Bank 2005). In Hong Kong (China) businesses cite traffic congestion, air pollution, and the need for a cleaner environment among the greatest obstacles to hiring international specialists and managers (AmCham 2006).

Public policies are a key tool for addressing current and future environmental degradation and natural resource use. How can policy makers and the public identify cost-effective opportunities for improving human welfare? How are public policies designed? Can they be designed in a way that is conducive to both economic growth and environmental sustainability? What is needed to allow them to do so? Does everyone benefit from a focus on environmental sustainability, or do some groups lose out? Does designing sustainable policies require difficult trade-offs and political maneuvering, or are such policies likely to be embraced by all?

This volume explores methodologies for addressing these questions. It focuses on Strategic Environmental Assessment (SEA) for integrating environmental considerations into policies, exploring how SEA can be used to design sustainable policies. The book does not provide recipes for designing sustainable policies or delve into the myriad meanings of environmental sustainability. It interprets *sustainable development* as economic growth that is socially equitable and environmentally responsible.[2] The term *environment* is used broadly to mean the biophysical environment as well as the linkages of the biophysical environment with the quality of life (health, livelihoods, and vulnerability) and economic activity (World Bank 2001). Hence environment is as much about the biophysical environment as about the people affected by environmental degradation. Paraphrasing an indigenous leader from Colombia, environmental sustainability is a state of being in which both environment and people are in harmony (Stephens 2006).

Evolution of SEA

SEA extends the application of environmental impact assessment (EIA) from projects to policies, programs, and plans.[3] National, regional, and international

SEA legislation usually falls under EIA legislation, extending its use to programs, plans, and, in some cases, policies. For example, national legislation in China requires SEAs of plans; the regional European SEA Directive (Directive 2001/42/EC) requires that SEAs be conducted for all programs and plans. Other countries also require SEAs for policies (as discussed later in this chapter).

Many SEAs (including those not driven by legislation) use a continuum of approaches rather than a single approach.[4] OECD (2006: 17) describes SEA as a "family of approaches using a variety of tools rather than a single, fixed and prescriptive approach." At one end of the spectrum, impact-based SEA integrates biophysical environmental considerations into higher levels of decision making by predicting potential effects of policies, plans, and programs on the environment and adopting the corresponding protection and mitigation measures. At the other end of the spectrum, institution-centered SEA aims to mainstream the environment and sustainability across higher levels of policy making by assessing the capability of the institutional and policy framework to detect environmental risks and its capacity to manage them in a timely and effective manner.

When SEA centers on impact assessment, it can be defined as "a systematic process for evaluating the environmental consequences of proposed policy, plan or programme initiatives in order to ensure that they are fully included and appropriately addressed at the earliest appropriate stage of decision making on par with economic and social considerations" (Sadler and Verheem 1996: 27). This definition (or variations of it) reflects an extension of "the EIA tradition and environmental concerns 'further up the chain' of decision making toward programmes and plans arising and dealt with in existing agencies and processes" (Connor and Dovers 2004: 165).

At the other end of the spectrum is the conceptualization of SEA "as a mechanism for mainstreaming environment and sustainability across the higher levels of policy making. . . . [It] suggests inadequacies of existing policy processes and thus a more substantial degree of organization and institutional reforms" than the concept of SEA as an upward extension of EIA (Connor and Dovers 2004: 165). This concept is reflected in a description of SEA as a participatory approach for upstreaming environmental and social issues to influence processes for development planning, decision making, and implementation at the strategic level. Implicitly included in this description is the importance of analytical work to support the decision-making process (Ahmed, Mercier, and Verheem 2005).[5]

Impact-Based SEA

Impact-based SEA evolved out of the field of EIA; many SEA advocates were EIA specialists who believed that assessments at the level of policies, plans, and programs could overcome the limitations of assessments conducted for individual projects (Thérivel and Partidário 1996). The terminology and procedures used for impact-based SEAs have counterparts in the EIA literature (box 1.1).

BOX 1.1
Impact-Based SEA Procedures

The methodology for impact-based SEA, and EIA, involves the following steps:

- *Screening.* Screening refers to the determination of the need for an SEA. If a proposed program, plan, or policy has a significant environmental impact, an SEA should be conducted.
- *Scoping.* Scoping refers to the identification of the impacts the SEA should assess. The scope of work (terms of reference) is usually determined by experts; in some jurisdictions the public is invited to participate in scoping.[a]
- *Identification, prediction, and evaluation of impacts.* The process of forecasting and evaluating impacts of programs, plans, or policies in an SEA can employ some of the same tools and procedures used in project-level EIA.[b] As in EIA work, professional judgment often plays a major role.
- *Mitigation.* Mitigation measures are intended to avoid, reduce, or offset the adverse effects of an action, such as the decision to approve a program or implement a plan.
- *Monitoring.* Monitoring the effects of plans or programs can alert the authorities to unintended outcomes that can be controlled by mitigation measures. By comparing predicted outcomes with those observed through monitoring, analysts may be able to improve their ability to predict impacts.

a. Canada's "Guidelines for Implementing the Cabinet Directive on the Environmental Assessment of Policy, Plan and Program Proposals" offers the following general advice on what should be included in the scope of work: "A strategic environmental assessment generally addresses the following five questions: (i) What are the potential direct and indirect outcomes of the proposal? (ii) How do these outcomes interact with the environment? (iii) What is the scope and nature of these environmental interactions? (iv) Can the adverse environmental affects be mitigated? and (v) Can positive environmental effects be enhanced? What is the overall potential environmental effect of the proposal after opportunities for mitigation has been incorporated?" (CEAA 2004a, Section 2.3).
b. The International Association for Impact Assessment uses public participation as a performance criterion for characterizing a "good-quality" SEA (IAIA 2002). For a review of techniques used for predicting and evaluating impacts in the context of SEA, see Thérivel (2004).

Applying SEA to Plans and Programs

Policies, plans, and programs are often viewed as forming a hierarchy, with policies at the top level, plans one level down, and programs at the lowest level. Programs make plans more specific by including details on an array of projects.[6]

Considerable experience exists in applying SEA to investment programs and plans.[7] National governments have conducted SEAs for investments programs. The SEA for the Argentina flood protection program, for example, assessed the cumulative effects of 50 individual flood protection subprojects in three river systems and identified the need for a component to improve coordination between cities and agencies in the flood plain—a provision that was included in a related project implemented in the 1990s (Garcia 1997).

At the program level, a successful SEA was carried out for power development options in the Nile Equatorial Lakes region. This SEA incorporated a multicriteria methodology to screen power development options, cumulative impacts assessment, mitigation plans, and power system planning in order to define an indicative least-cost regional power master plan for the subregion (World Bank 2007a).

The use of SEAs for watershed plans has been applied in developing countries (one example is the SEA for the Palar water basin in Tamil Nadu, India). This SEA uses both analytical and participatory processes to internalize environmental considerations in water resources planning in order to frame a common development vision for the basin (World Bank 2007b).

Applying SEA to Policies

In contrast to the pervasive application of SEAs to programs and plans, the application of SEAs to policies has been scarce. For example, the EU SEA Directive, while requiring SEA for all programs and plans, does not mention SEA for policies. However, a number of countries, including Canada, Denmark, the Netherlands, and New Zealand, have implemented procedures to incorporate environmental consideration into the design of public policies (table 1.1). Legislation on SEA of policies exists in some developing countries as well (including the Dominican Republic and Kenya), but implementation is rare. Such legislation typically extends impact-based SEA methodologies to policy application (Ahmed and Fiadjoe 2006).

TABLE 1.1
Examples of SEAs for Policies

Country	Instrument
Canada	Policy Impact Assessment is applied to appraise environmental effects of policies and cabinet-level decisions (CEAA 2004a).
Denmark	SEA is applied on bills and other proposals likely to have significant impacts on the environment (Sadler and Verheem 1996).
Finland	The Norm Law, issued in 1996, requires application of SEAs to policies.
Netherlands	Environmental Test (E-Test) aims to assess the environmental effects of policies.
New Zealand	Policy statements and plans must be evaluated to determine if the goal of the Resource Management Act (1991), which promotes sustainable management of natural and physical resources, is achieved (Sadler and Verheem 1996).

Source: Authors.

While it still has to enter force, the Kiev (SEA) Protocol associated with the Convention of Environmental Impact Assessment in a Transboundary Context (the Espoo Convention) also mandates the application of SEA to programs and plans required under the protocol. However, in contrast to the EU SEA Directive, it includes a softer reference as applied to policies. Specifically, it states that "each party shall endeavor to ensure that environmental, including health, concerns are considered and integrated to the extent appropriate in the preparation of its proposals for policies and legislation that are likely to have significant effects on the environment, including health."

Toward an "Institution-Centered" SEA Approach

This volume explores how to improve the effectiveness of SEA application to policies and presents a new conceptual and methodological framework for applying SEA to policies.

Initially, in chapter 2, Ortolano assesses several applications of SEA to policies. Canada and the Netherlands have extensive experience conducting environmental assessments for policies, plans, and programs. However, the agencies required to prepare these assessments have often done so on a pro forma basis, as Ortolano shows in chapter 2. Performance was perfunctory partly because assessments often occurred late in the policy formulation process and partly because agencies could often marginalize environmental assessment requirements without penalty. Key factors affecting policy SEA in Canada and the Netherlands include difficulties in applying impact assessment methodologies to policy proposals, the absence of cross-sectoral agencies with responsibility for overseeing compliance with the SEA requirements, and the lack of commitment to SEA by top-level agency officials.[8] The case study analysis in chapter 2 shows that the influence of policy-level SEA on policy formulation and implementation depends heavily on process-integration issues, especially when SEA begins relative to the policy-formulation and implementation process and how often SEA teams and policy designers interact.

In chapter 3, Feldman and Khademian argue that the application of impact-based SEA has had a limited impact on public policy because it is built on an understanding of policy formation as occurring in linear stages of rational decision making. In reality, decisions are made in complex settings in which preferences are unclear, technologies or the means of getting things done are not known, and participation in the process is fluid. Feldman and Khademian present two models that acknowledge the complex context of policy formation as a continuous process. One, "adaptive management," is oriented to enabling action in the face of uncertainty; the other, "inclusive management," is oriented to enabling action in the face of ambiguity by encompassing more viewpoints in policy formation, including those that have been marginal in other processes.

The daunting and politically difficult task of setting environmental priorities in a policy decision-making process is the first SEA element in devising effective and cost-effective strategies for addressing environmental problems. Based on experiences with the application of environmental priority-setting techniques in developing and developed countries, in chapter 4 Morgenstern examines alternative approaches to environmental priority setting that emphasize quantitative techniques. These techniques evaluate risks and economic damages, in particular through comparative risk and economic damage assessments.

Several studies show that giving the most vulnerable a voice helps policy makers understand the synergies between environmental goals, economic growth, and poverty reduction. Thus a pillar of an institutions-centered SEA should be giving the vulnerable greater voice in policy formulation, especially where environmental considerations are involved. In chapter 5 Kende-Robb and Van Wicklin explain the importance of ensuring that decision makers not only seek the voice of the vulnerable but also act on it—by creating "space" for public participation, identifying entry points in the policy process for increasing voice, defining different levels of participation, and using tools for amplifying the voice of the vulnerable and making sure that voice is heard in the policy process.

To craft and nurture sustainable policy initiatives that can address externalities in ways that will have positive impact on the environment, SEA processes need to support long-term constituencies that want to support such policies and can hold policy makers accountable for their performance in implementing them. Transparency is critical in the policy process needed to allow these constituencies to demand accountability from policy makers. In chapter 6 Blair constructs a theoretical framework that fits accountability, transparency, and long-term constituency building together as part of the policy process. He illustrates this framework with case studies from Delhi and Indonesia.

In chapter 7 Ebrahim argues that policy learning should not be viewed as a rational and technocratic process. Instead, learning should emphasize the political and institutional contexts within which opportunities for policy learning emerge. Opportunities for policy learning appear at different stages: agenda- or priority-setting on environmental issues, stakeholder access and representation in policy formulation, and accountability in implementation.

Drawing on the lessons emerging from chapters 2–7, as well as case studies documented elsewhere, in chapter 8 Ahmed and Sánchez-Triana propose a new conceptual and methodological framework for applying SEA to policies. This "institution-centered SEA" is being piloted by the World Bank in different regions of the world. The authors suggest that SEA can be used to help design and implement equitable and environmentally sustainable policies by adding a third pillar designed to enhance learning and continuous improvement of policy design and implementation to the existing two analytical and participatory pillars in traditional SEA methodology. This approach takes into account the complex process of policy formation and the importance of seizing opportunities for policy reform as they arise. Elements within the approach also seek to enhance the creation of windows of opportunities for future policy reform.

Notes

1 In Cambodia about 80 percent of the population lives off of agriculture, forest, and fish resources (Evans 2006).

2 Ever since the release, in 1987, of the report of the World Commission on Environment and Development (the Brundtland Commission), there has been emphasis on the importance of sustainable development, a concept that weaves economic growth, environmental protection, and social justice as complementary goals across generations. The 1992 Rio Earth Summit reemphasized this message. The Millennium Development Goals—particularly goal 7, target 9, which strives to "integrate the principles of sustainable development into country policies and programs and reverse loss of environmental resources"—also reiterate it.

3 Examples of the limitations of EIA that can be overcome by SEA include its inability to account for the cumulative effects of multiple, successive projects in a particular area or to focus attention on strategic choices that, had they been made, would have precluded the need for the project considered in the EIA (see Thérivel and Partidário 1996; Connor and Dovers 2004). A 2004 intergovernmental policy forum on environmental assessment characterized as "core premises . . . that SEA will lead to fewer and/or simpler EIAs and will be more effective in identifying issues of cumulative impact" (CEAA 2004b: 17).

4 See Dalal-Clayton and Sadler (2005) for a comprehensive review of SEA.

5 This description of SEA as consisting of both analytical and participatory approaches is consistent with the most recent use of the term *SEA* by the OECD Development Assistance Committee Network on Environment and Development Cooperation (ENVIRONET). The latter describes SEA as "analytical and participatory approaches to strategic decision-making that aim to integrate environmental considerations into policies, plans and programmes and evaluate the inter linkages with economic and social considerations" (OECD 2006: 17).

6 Of course, real systems are often more complex than this hierarchy suggests (Dalal-Clayton and Sadler 2005).

7 Dalal-Clayton and Sadler (2005) document the extent of SEA activity around the world. Donors are trying to harmonize approaches to SEA and to identify new opportunities for its application. The Organisation for Economic Co-operation and Development's recent "Good Practice Guidance on Applying SEA in Development Cooperation" (OECD 2006) is one such example. The World Bank's application of SEA initially arose from a policy requiring environmental assessment in all investment projects (that policy also provided for the use of sectoral or regional environmental assessment in certain contexts). In 1999 the requirement was extended to sectoral adjustment loans, for which SEA was often the tool of choice. The World Bank's 2001 Environment Strategy recognized SEA as a key means of integrating environment into the sectoral decision-making and planning process at early

stages and made a strong commitment to promote the use of SEA as a tool for sustainable development.

8 This assessment was carried out in 2005, relying on the information available at the time. It does not take into account any changes to SEA systems made after June 2005.

References

Ahmed, Kulsum, and Yvonne Fiadjoe. 2006. *A Selective Review of SEA Legislation: Results from a Nine Country Review.* Environment Strategy Paper 13, World Bank, Washington DC.

Ahmed, Kulsum, Jean-Roger Mercier, and Rob Verheem. 2005. *Strategic Environmental Assessment: Concept and Practice.* Environment Strategy Note 14, World Bank, Washington, DC.

AmCham (American Chamber of Commerce in Hong Kong). 2006. "Polluted Air Threatens Business Decline in Hong Kong." *Journal of the American Chamber of Commerce in Hong Kong* 38 (8): 9–16.

CEAA (Canadian Environmental Assessment Agency). 2004a. "Guidelines for Implementing the Cabinet Directive on the Environmental Assessment of Policy, Plan and Program Proposals." Updated February 11. Ottawa. http://www.ceaa.gc.ca/016/directive_e.htm#2.

———. 2004b. "Status of Progress and Emerging Challenges in EIA and SEA: Ten Years after the Quebec Summit." Ottawa.

Connor, R., and S. Dovers. 2004. *Institutional Change for Sustainable Development.* Cheltenham, United Kingdom: Edward Elgar.

Dalal-Clayton, B., and B. Sadler. 2005. *Strategic Environmental Assessment (SEA): A Sourcebook and Reference Guide to International Experience.* London: Earthscan.

Evans, Patrick. 2006. "Flood Forests and Community Fisheries on the Tonle Sap Great Lake, Cambodia." *European Tropical Forest Research Network.* http://www.etfrn.org/etfrn/newsletter/news4546/.

Garcia, Luis. 1997. *Evaluacion ambiental del programa de control de inundaciones.* Consultant report, La Plata, Argentina.

IAIA (International Association for Impact Assessment). 2002. *Strategic Environmental Assessment Performance Criteria.* Special Publication Series 1, Fargo, ND. www.IAIA.org.

IUCN (World Conservation Union). 2006. "Sustainable Livelihoods." http://www.iucn.org/en/news/ archive/2001_2005/mbsustliveli.pdf.

North, Douglass C. 1990. *Institutions, Institutional Change and Economic Performance.* Cambridge: Cambridge University Press.

OECD (Organisation for Economic Co-operation and Development). 2006. *Applying Strategic Environmental Assessment (SEA): Good Practice Guidance for Development Co-operation.* DAC Guidelines and Reference Series. Paris: OECD Publishing.

Pruss-Ustun, A., and C. Corvalan. 2006. *Preventing Disease through Healthy Environments: Towards an Estimate of the Environmental Burden of Disease.* Geneva: World Health Organization.

Sadler, B., and R. Verheem. 1996. *Strategic Environmental Assessment: Status, Challenges and Future Directions.* Publication 53, Ministry of Housing, Spatial Planning and the Environment, the Hague.

SECTUR (Secretaria de Turismo). 2002. *Programa nacional de turismo 2001–2006: El turismo la fuerza que nos une,* 2nd. ed. Mexico City.

Stephens, Carolyn. 2006. "Integrating Environmental Health into Poverty Reduction Strategies and Sustainable Development." Paper presented at the Ninth Poverty-Environment Partnership Meeting, Washington DC, June.

Thérivel, R. 2004. *Strategic Environmental Assessment in Action*. London: Earthscan.

Thérivel, R., and M. R. Partidário. 1996. *The Practice of Strategic Environmental Assessment* London: Earthscan.

UNDP (United Nations Development Programme). 2004. *Reducing Disaster Risk: A Challenge for Development*. New York: UNDP Bureau for Crisis Prevention and Recovery.

———. 2007. "How Tourism Can Contribute to Environmental Conservation." http://www.uneptie.org/pc/tourism/sust–tourism/env–conservation.htm.

World Bank. 2001. *An Environment Strategy for the World Bank*. Washington DC.

———. 2005. *Mexico Second Programmatic Environment Development Policy Loan*. Program Document, Report 32249–MX, Washington, DC.

———. 2006. *IDA's Thirteenth Replenishment: A Retrospective Review*. Report 36321, Board Operations, Washington, DC.

———. 2007a. *Strategic Sectoral, Social and Environmental Assessment of Power Development Options in the Nile Equatorial Lakes Region*. Report No. 39199, Africa Region, Nile Basin Coordination Unit, Washington, DC.

———. 2007b. *Strategic Environmental Assessment and Integrated Water Resources Management and Development*. Economic and Sector Work, Environment Department, Washington, DC.

C H A P T E R   2

Policy-Level Strategic Environmental Assessments: Process Integration and Incentives of Policy Proponents

Leonard Ortolano

COUNTLESS ENVIRONMENTAL IMPACT ASSESSMENTS (EIAs) have been completed for individual projects, and many hundreds of strategic environmental assessments (SEAs) have been prepared for areawide and sectoral plans and programs. Fewer EIAs have been produced for policies than for plans and programs.[1] Sadler (2005) lists about a dozen countries with well-documented experiences with policy-level SEAs.[2] Interest in policy-level SEAs has mounted rapidly, and efforts have been made to extract lessons from experience (Sadler 2003).[3]

Even if a new environmental assessment process intended to influence policy design is created, it does not follow that those who receive the results of the assessments use them in ways that enhance policy making. This chapter argues that two considerations are of overriding importance in determining the influence of policy-level SEAs on policy designs: the mode of integrating SEA into policy-making

Leonard Ortolano is the UPS Foundation Professor of Civil Engineering at Stanford University. Jiri Dusik, of the Regional Environmental Center for Central and Eastern Europe, in Szentendre, Hungary, had an important influence on this chapter; the author is very grateful to him for sharing some of his many experiences in conducting SEAs. The author would also like to thank Kulsum Ahmed, Rachel McCormick, Ernesto Sánchez-Triana, Jaye Shuttleworth, and two anonymous reviewers for their comments and suggestions on early drafts of sections of this chapter.

processes (which Dalal-Clayton and Sadler 2005 refer to as *process integration*) and the incentives policy proponents have to consider the results of SEAs.

In an effort to synthesize lessons from experience with policy-level SEAs, this chapter examines five case studies of policy-level assessments intended to affect policy formulation. It also analyzes evaluations of national requirements calling for environmental assessments of proposed laws and policies in Canada and the Netherlands.[4] Examination of each of these types of policy-level SEAs provides insight into two central questions: How can SEA processes be structured to allow SEA teams to interact with policy designers in ways that facilitate the consideration of the environment in policy making? What steps can be taken to provide policy designers with incentives to use environmental assessment results in policy making?

Because the environmental assessment studies considered here have different objectives (from formulating a national energy policy to drafting legislation) and occur in diverse contexts and cultures, many dimensions are not comparable. In all cases, however, designers of the SEA studies or the SEA program requirements were required, at least implicitly, to deal with both of the questions posed above. The responses to this common set of questions are comparable in a general sense, and they provide a basis for drawing some tentative conclusions that can be useful in SEA program design.

Case Studies

The synthesis of five case studies of policy formulation presented here provides a close look at policy-level SEA in practice. Four of the five cases are based on material prepared by Environmental Resources Management, an environmental consulting firm (ERMs 2004). The other case, which concerns water and sanitation in Argentina, was reported in an unpublished paper by World Bank staff. For three of the five cases (those involving Argentina, Canada, and the Czech Republic), supplemental information was obtained from participants in the original SEA work, and wherever possible, from supplemental literature not cited in the original SEA case studies.

The synthesis focuses on the extent to which SEA processes were integrated into policy-making processes and influenced policy designs. (In this context, "design" includes provisions for monitoring policy outcomes.) It indicates where SEAs enhanced intersectoral coordination and public involvement; because this coordination and public involvement influenced how policy makers considered the environment.

An issue not examined here concerns how environmental outcomes observed on the ground were affected by proposed policies created based on recommendations in SEAs. The reason for this exclusion is that the ERM and World Bank documentation was for ex ante SEAs (that is, assessments conducted before decisions are made). In order to determine how an SEA influenced outcomes of policy implementation, it would be necessary to conduct ex post studies and to make causal arguments linking actions taken in response to ex ante SEAs with observed outcomes.[5]

Argentina: Policy Reform in Water and Sanitation Sectors

During the late 1990s, the government of Argentina requested World Bank assistance in reforming its water and sanitation sectors.[6] The government's proposal concerned policy reforms for medium-size cities and issues linked to earlier reforms in larger cities. The Bank asked the government to prepare an environmental assessment for what was termed the Argentina Water Sector Reform Project. In addition to policy reforms, the project concerned measures to finance public works, such as municipal water and wastewater treatment facilities. The government hired consultants to prepare an SEA focusing on linkages between Argentina's environmental regulatory activities and its water and sanitation sectors. The undersecretary of water resources prepared the terms of reference (TOR) for the consultants.

The SEA was conducted during project preparation. In a departure from a typical SEA approach, the government's TOR required the consultants to identify priority issues by examining "negative external effects" (that is, unaccounted-for costs imposed on others) in connection with: excessive water losses, such as leaks in water distribution systems; water quality deterioration, such as degradation caused by pollution; and adverse effects of water and waste treatment systems, such as noise and odors.

The SEA was carefully integrated into the process of designing water and sanitation sector reforms. SEA consultants coordinated with relevant policy-design authorities in Argentina, particularly the Ministry of Economy. Interestingly, the ministry initially had reservations about the SEA, because it felt EIAs for individual projects could satisfy World Bank safeguard requirements. However, the ministry became a supporter of the SEA after the consultants showed how a fragmented legal framework and weak enforcement of environmental regulations contributed to water quality degradation, which interferes with provision of potable water. The ministry's change in attitude from reluctance to enthusiasm about SEA demonstrates that policy makers who see how SEA advances their agencies' missions may become SEA advocates.

In framing their analysis, government officials, World Bank task team members, and SEA consultants visited several cities and requested information from local water authorities and other stakeholders. These visits, along with several workshops, helped identify regulatory issues blocking expansion of water service provision and participation by private water operators. The consultants traveled to some of the poorest neighborhoods in order to learn about the needs of vulnerable populations.

Loan conditions recommended in the SEA were incorporated into the final policy reform arrangements between the World Bank and the government of Argentina. The project included an environmental institution-building component, which represented about 2 percent of total project cost. These funds were to finance, among other things, revisions of environmental standards, preparation of EIA guidelines, and enhancement of institutional capacity to manage watersheds. The SEA process also helped government officials move from a narrow focus on environmental impacts of individual projects to a broader concern with sector-level priorities and environmental management issues that could be addressed only through policy reforms. The SEA recommended information-gathering activities and other measures to foster monitoring of progress.

Canada: SEA for NAFTA

A 1990 Canadian cabinet directive (updated in 1999 and 2004) requires SEAs for proposed policies, plans, and programs before they are brought before the cabinet. A few years before issuance of the 1993 guidelines implementing this directive, the Canadian government undertook an environmental review of the North American Free Trade Agreement (NAFTA). The resulting document, "North American Free Trade Agreement: Canadian Environmental Review," is viewed here as an SEA.

NAFTA negotiations took place between June 1991 and August 1992. The final draft SEA was published in October 1992, when the text of NAFTA was initialed by trade negotiators for Canada, Mexico, and the United States, enabling the countries to work toward domestic approval of the draft treaty.

The Canadian SEA for NAFTA did not reflect an effort to fully integrate activities related to SEA into the process of drafting NAFTA, because the SEA was conducted after negotiations had started.[7] Production of an SEA was the last, and arguably the least-important, mechanism for integrating environmental considerations into NAFTA's design. The following four-point plan was used for this purpose (ERM 2004).

- Environmental representatives were appointed to the International Trade Advisory Committee (ITAC) and each of the 15 Sectoral Advisory Groups on International Trade (SAGITs).
- Trade-related environmental concerns were integrated into all phases of NAFTA negotiations.
- Parallel discussions were initiated on environmental cooperation by Canada, Mexico, and the United States.
- The NAFTA Environmental Review Committee conducted an SEA.

ITAC and the SAGITs were the principal mechanisms for public involvement. These groups included representatives of business, the environment, labor, and academia; results of consultations with ITAC and the SAGITs were reported to the Trade Minister (Hazell and Benevides 2000). In February 1992 the chairs of ITAC and the SAGITs, along with environmental representatives and Canada's senior NAFTA negotiators, met to discuss environmental issues. A similar group met in April 1992 for a workshop on trade and environment.

The SEA process included provisions for interagency coordination. The NAFTA Environmental Review Committee—led by the Department of Foreign Affairs and International Trade (DFAIT)—included representatives from many federal departments (Shuttleworth 2005). The committee consulted with Canadian negotiators, provincial officials, and ITAC and the SAGITs; it also held workshops for environmental groups. An extensive stakeholder consultation process involved government ministers, environmental agencies, and university professors (ERM 2004).

The TOR for the Canadian NAFTA SEA was made available to the public, and citizens submitted comment letters.[8] The general public did not participate directly in treaty design, however.

One of the main concerns of the environmental review was that treaty ratification might reduce Canada's ability to regulate environmental quality (Shuttleworth 2005). The review concluded that this would not be the case.

Documentation does not allow the factors that led to either specific environmental provisions in NAFTA or the treaty's environmental side-agreements to be disentangled. NAFTA's negotiators considered environmental concerns, and the NAFTA Environmental Review Committee played a role in raising environmental concerns with the Canadian negotiators, but the SEA process was only one element of a four-part plan to integrate environmental concerns into treaty design. Moreover, no evidence links a particular provision in either NAFTA or the environmental side-agreements to the process of producing the SEA document. It is unlikely that the SEA played a major role in negotiations, because treaty negotiators learned of environmental concerns through the earlier work of the ITAC and SAGITs.

Environmental controversies surrounded the creation of NAFTA, and provisions were made to monitor policy outcomes linked to the environment. A committee on standards-related measures was formed to enhance environmental standards and cooperation among member states, and the North American Commission on Environmental Co-operation was created to assess ongoing impacts of NAFTA implementation.

In 2002 Foreign Affairs Canada established a new procedure for conducting SEAs for trade agreements.[9] The new approach is similar to the Argentine case, because the SEA team works independently and in parallel with treaty negotiators before and during the negotiation process (J. Shuttleworth, personal communication, June 1, 2005). It resembles the Slovak process (described next) in that a member of the SEA team is also part of the treaty negotiation group, which allows negotiators to become aware of SEA results as they become available.

The Czech Republic: Tourism Policy

In 1992 the government of the Czech Republic enacted the Czech Environmental Impact Assessment Act (No. 244/1992).[10] Section 14 of the act requires environmental assessments for "development concepts" submitted to or approved by central administrative authorities in several sectors, including tourism. (The term *concept* is widely understood to refer to strategies, policies, plans, and programs adopted by public authorities.)

Under Section 14 the proponent of a development concept must prepare SEA documentation, including elements typically contained in project-level EIAs. The proposed concept and the associated SEA documentation are subject to public review using arrangements determined jointly by the proponent and the Ministry of the Environment. After public review, the proponent forwards review comments along with the concept and the SEA documentation to the Ministry of the Environment, which then issues an "SEA standpoint." Although the proponent need not accept the standpoint's recommendations, the government will not approve a concept unless it is accompanied by a standpoint.

Use of SEAs in the Czech Republic was limited before 1996. Indeed, some central governmental organizations evaded SEA requirements by changing the names of documents for programs and policies so they would not be viewed as "concepts."

In a departure from requirements in Section 14, the Ministry for Regional Development created Tourism Policy 2000 without preparing an environmental assessment; the government approved that policy in preliminary form. Soon thereafter the ministry prepared its draft Sectoral Operational Program for Tourism and the Spa Industry (SOP), which was also created without a required SEA. After the Ministry of the Environment advised the government that SEAs were required, the government decided to withhold its final approval of the tourism policy until after an SEA was completed and the policy was resubmitted.

Initially, the Ministry for Regional Development had reservations about preparing the SEAs, fearing that bureaucratic hurdles would slow a process it viewed as nearly complete. The Regional Development Ministry's lack of eagerness to conduct SEAs is reflected in the Ministry of the Environment's need to intervene to have the SEAs performed. There was a delay of several months in starting the SEAs because of discussions between the two ministries.[11]

The SEA process unfolded in an unorthodox fashion. Under normal circumstances, an SEA would have been conducted for the proposed tourism policy, and the final policy would have provided a basis for the SOP, which would have had its own SEA. In this instance, when the draft SOP was nearly completed, the Ministry for Regional Development asked the Ministry of the Environment whether an SEA was needed.[12] The Ministry of Environment responded in the affirmative; the Regional Development Ministry then engaged consultants to prepare an SEA for the SOP. Once that assessment was completed, work began on the SEA for the tourism policy.

The public participated in each of the SEA processes. For the SOP, five workshops were held in different regions. The SEA process for the tourism policy also allowed for public involvement, through a national public hearing. In both SEAs public comments were positive.

Although both SEAs were conducted after the Ministry for Regional Development had, in its own view, completed much of its policy design work, they led to notable changes in the tourism policy and SOP. These changes included: addition of new policy goals (including "support of environmentally friendly mass transport at more popular destinations"); environmental targets for projects that would be approved under the SOP (including "reduction of excess of visitations [at] the most heavily visited destinations"); and new measures (including "measures to introduce a system of [accreditation of] tourism destinations") (ERM 2004: 26–27). Interestingly, the ministry eventually became enthusiastic about the SEA.

According to Jiri Dusik, a member of the team of consultants that prepared the SEAs, the influence of the SEAs on the tourism policy and SOP can be explained in two parts. The first concerns how vigorous debates between the SEA team and the staff of the ministry were facilitated.[13] Dusik gives particular credit to one of his co-team members, an environmental management specialist with a talent for reducing tensions in heated discussions.

The second concerns the structure of the SEA process. After first developing an understanding of the entire policy formulation process, the team sequentially assessed each of the four policy dimensions: overall context, goals and objectives, proposed measures, and implementation arrangements. The team would not assess any particular dimension without first completing its work on the preceding ones (for example, it would not analyze proposed measures until it had completed its work on context and objectives).

When the SEA process began, the Ministry for Regional Development felt the team would eventually try to discredit its proposed measures and activities; discussions of the team's work on overall context were animated. After the SEA team assessed goals and objectives set out in the SOP, the team met with 15 representatives of different departments within the ministry. The team's recommendations for changes in the goals section of the SOP were seriously considered and extensively debated. After two days of meetings, the SEA team's recommendations for changes in goals were accepted. This activity was eventually repeated for the tourism policy, with the ministry readily accepting recommended changes.[14]

By the time discussions concerning goals had been completed, the ministry understood the overall direction being followed by the SEA team and trust had developed between the team and ministry staff. When the team's meetings with the ministry on proposed measures and activities in the SOP began, solid working relationships had been established and the groundwork for productive negotiations laid. During these meetings the SEA team outlined possible impacts and proposed mitigation measures for the SOP, which were discussed over a three-to four-day period. Many of the team's recommendations were eventually accepted. Corresponding changes in the tourism policy were later agreed on in less than a day.

Interactions between the SEA team and the ministry regarding recommendations for implementation and monitoring took place at a workshop, at which indicators were created for use in monitoring progress toward goal attainment. Although the Ministry for Regional Development had agreed to conduct monitoring activities, staff turnover hindered the progress on monitoring.[15] Between 2002 and 2005, nearly all of the staff members in both the Tourism Department of the Ministry for Regional Development and the SEA Department of the Ministry of the Environment were replaced with new staff. This slowed the monitoring effort, because most staff members who agreed to conduct monitoring had left the ministries before a final, detailed monitoring plan had been created, and replacement staff were not well positioned to follow up. Some monitoring has been conducted, but data gathering has not been as systematic as originally envisioned.

An important aspect of the SEA process concerns the learning that has taken place within the Ministry for Regional Development. The assessment process itself was transformative: by the time the SEA was completed, ministry staff who had begun the process with significant misgivings about SEA had become enthusiastic about it, suggesting that policy makers may embrace the process once they learn what SEA is about and how it can help meet the goals of their agencies.[16]

Interestingly, individuals who replaced members of the Tourism Department who had left following completion of the assessment are also supporters of SEA (J. Dusik, personal communication, January 17, 2006). In January 2006 the Ministry for Regional Development initiated a process to update the Czech tourism policy; on its own initiative, Tourism Department staff decided to conduct an SEA simultaneously with the update of the tourism policy. The TOR for the proposed SEA is comprehensive and specifies the stages of policy formulation at which inputs from the SEA process are expected. In addition, the new SEA is intended to yield a monitoring plan detailed enough to be implemented by the ministry. In this way, the ministry hopes to avoid the difficulties caused by the lack of detail in the monitoring plan for the initial version of the tourism policy.

The Slovak Republic: Energy Policy 2000

Under the Slovak Republic's EIA Law, initially implemented in 1994, environmental assessments are required for development policies in a number of sectors, including energy.[17] The law requires the ministry designing a policy to prepare a draft policy reflecting environmental considerations and to inform the public of the draft at least two months before the Ministry of the Environment reviews it. The proposing ministry must then confer with the Ministry of the Environment before submitting the policy for government approval.

The SEA process was indistinguishable from the process employed to design Energy Policy 2000.[18] Integration of environmental considerations into policy making was part of the eight-step process used (ERM 2004). This process included the following steps:

- Preparation of draft policy
- Public notification of preparation of policy
- Formal consultations and public participation
- Public hearing on draft policy
- Statement by Ministry of Environment on draft policy
- Revision of draft policy
- Adoption of final policy
- Monitoring of policy implementation

The way in which SEA is integrated into the policy formulation process has "probably [been] the most important factor behind the effectiveness of Slovak SEAs" (ERM 2004: 44).

Using this process the proposing ministry and the Ministry of the Environment coordinate closely when the Ministry of the Environment prepares its statement on the draft policy and when the draft policy is revised. The proposing ministry is not obligated to accept recommendations in the Ministry of the Environment's statement; revisions are made by mutual consent. In designing Energy Policy 2000, the proposing ministry (the Ministry of Economy) revised its draft policy based on results of the SEA process. This represented an improvement in interagency coordination (ERM 2004).

Public involvement was a hallmark of the SEA for Energy Policy 2000. The SEA provided a vehicle for mobilizing NGOs concerned with energy-environment relations. The Ministry of Economy circulated a preliminary draft of the energy policy to NGOs for comment before finalizing its draft policy. The draft was included on several government, university, and NGO Web sites, and the public was notified of the draft through newspaper announcements. Media coverage was extensive, the draft policy was made available at district and regional governmental offices, and an NGO established information kiosks in several towns. In addition, interested parties used the Internet to exchange information. By the end of the public comment period, the Ministry of the Environment had received hundreds of comments.

NGOs organized themselves under an umbrella organization, Energy 2000, and formulated their own proposal, the New Energy Policy of the Slovak Republic. This proposal was posted on government and NGO Web sites and discussed extensively. Energy 2000 also organized an international conference, at which both the new energy policy and the Ministry of Economy's draft policy were debated. A public hearing on the government's draft policy and the new energy policy (organized by the Ministry of the Environment and the Ministry of Economy) drew about 150 participants. Transcripts of the hearing, government consultations with experts, and several hundred public comments formed a basis for the Ministry of the Environment's formal statement, which was made available to the public on request.

The SEA process, which considered economic as well as environmental and social factors, yielded many modifications to the draft energy policy. For example, the government's draft was changed to encourage diversification of energy sources and "de-monopolization" and decentralization of the energy sector. Although monitoring the environmental impacts of implementing Energy Policy 2000 was not legally required, the Ministry of Economy committed to a monitoring program. Monitoring was conducted by "responsible government bodies," and the government was following through on commitments made in the SEA (Mária Kozová, Comenuius University, Bratilslava, Slovak Republic, personal communication cited in ERM 2004: 40).

South Africa: KwaZulu-Natal Trade and Industrial Development Policy

South Africa's 1995 Development Facilitation Act provides the foundation for addressing environmental concerns in the context of spatial planning, but it does not require SEAs. Notwithstanding the absence of a legal mandate, several jurisdictions in South Africa have voluntarily undertaken SEAs (Wiseman 2000). Passage of the 1998 National Environmental Management Act enabled the Department of Environment and Tourism to issue guidelines for "integrated environmental management"; SEA has become a popular tool for implementing this management approach (ERM 2004; see also Rossouw and Wiseman 2004). In 2000 the Council of Scientific and Industrial Research (CSIR), in partnership with the Department of Environment and Tourism, issued a guideline document on SEA (CSIR 2000).

In response to the 1995 Development Facilitation Act, the KwaZulu-Natal Regional Economic Forum asked CSIR to prepare an SEA providing input for creating the forum's trade and industry policy for the KwaZulu-Natal region. The forum, made up of representatives of regional government, industry and NGOs, was responsible for creating this policy; it wanted information on types of development that would be possible (Wiseman 2000).

The SEA process was completed before the policy-making process started; "it is not clear how the SEA was integrated into policy, if at all" (Wiseman 2000: 161). Because the SEA was an input into a subsequent policy design process, it had no bearing on either intersectoral coordination or the monitoring of policy outputs. Apart from two stakeholder conferences involving members of the Regional Economic Forum, CSIR and its subcontractors conducted the SEA on their own.

The SEA was "mainly an analytical exercise, based on the spatial resources baseline—that is, data embedded in a geographic information system—overlaid with possible future industrial development scenario impacts" (ERM 2004: 46). It was completed during 1996, the same year the forum created its trade and industrial development policy.

CSIR began its SEA work with technical studies; it then held a stakeholder workshop to discuss key environmental issues. In preparation for subsequent stages of the study, the SEA team identified assessment criteria for the degree to which an industrial development scenario would lead to waste emissions in excess of "assimilative capacity" (the ability of a body of water or the atmosphere to receive waste without significant deleterious effects); the demands of the proposed scenario on use of natural resources; and the suitability of particular land areas for different types of industry.

In the final stage of its work, the team applied the criteria to classify the environmental issues and baseline conditions as strengths, weaknesses, opportunities, or threats in the context of particular industrial development scenarios. The SEA results included information on how baseline environmental conditions constrained development and how particular development scenarios would affect the environment.

After the CSIR team had conducted its analytic work, a second stakeholder meeting was held. Forum members commented on results; in response CSIR modified its SEA report. No apparent causal link exists between the SEA and the eventual trade and industrial policy issued by the forum. However, the industrial development sectors identified for development in the KwaZulu-Natal Trade and Industrial Development Policy were featured in the SEA (ERM 2004).

Analysis of Case Studies

The case studies highlight the importance of the link between policy design outcomes and the way SEAs are integrated into policy-making processes.[19] Several lessons can be drawn from these studies.

Results Viewed in the Context of Integration

The cases can be classified in terms of the four ways in which SEA integration was carried out (table 2.1).

The outcomes of the five cases studied can be assessed using this typology (table 2.2). The assessment reflects only general tendencies; case study documentation does not permit a more complete characterization. The mode of integration was readily determined by considering when, in the policy-design process, the SEA team interacted with the policy designers. In some cases interagency coordination mechanisms and public involvement programs were created as part of the SEA process. In contrast, it was challenging to determine whether SEA processes

TABLE 2.1
Integration of SEA into Policy Making

Level of Integration	Description	Example
Complete	SEA specialists are part of the policy design group; no clear distinction exists between policy making and SEA.	Energy policy in the Slovak Republic
Partial/simultaneous effort	SEA experts form a team that is distinct from the policy-making body; the team works cooperatively and in parallel with the policy-making group, with multiple points of contact.	Water and sanitation sector reforms in Argentina
Partial/late-stage effort	SEA team works in response to a draft policy proposal. Multiple points of contact may exist between SEA experts and the policy-making body, but contacts occur relatively late in the policy-design process.	NAFTA and the Czech Republic's tourism policy
Partial/technical support	SEA is conducted primarily to provide technical information to support policy formulation; integration occurs at a discrete point in the policy-design process.	Trade and industrial development policy in KwaZulu-Natal

Note: Other researchers have distinguished between full and partial SEA integration (see, for example, Dusik and Kosikova 2004).
Source: Author.

TABLE 2.2
Level of Integration of SEAs in Case Studies

Level of Integration	Complete — Slovak Republic	Partial/ Simultaneous Effort — Argentina	Partial/Late-stage Effort — Canada	Partial/Late-stage Effort — Czech Republic	Partial/Technical Support — South Africa
Intersectoral coordination	Yes	Yes	Yes	Yes	No
Public involvement	Yes	Yes	Yes	Yes	No
Influence on policy design	Yes	Yes	Unclear	Yes	Unclear
Influence on monitoring	Yes	Yes	Unclear	Unclear	No

Source: Author.

influenced policy design, except where specific recommendations in SEAs became part of final policies. It was particularly difficult to determine whether and how monitoring programs were influenced by SEAs, because the case study documentation often did not contain information on whether monitoring programs agreed to in SEA processes were carried out.

Several observations can be made based on this assessment:

- Even partially integrated SEAs can affect designs in significant ways (as the cases of Argentina and the Czech Republic reveal).

- The influence of SEA on coordination depends on opportunities the SEA team has to interact with technical experts in agencies and policy designers. The only SEA showing no discernable effects on improving intersectoral coordination was the one in South Africa.

- The degree of public involvement in SEA is highly variable, reflecting the different political and cultural contexts in which SEAs are conducted.

- SEA documentation sometimes provides information on whether an SEA process led to design of an ex post monitoring plan that was implemented: the Slovak SEA clearly affected monitoring, and the South African SEA did not. The Czech Republic SEA yielded a general agreement on monitoring activities, but changes in staffing at the relevant ministries slowed implementation of the monitoring effort.[20]

- The potential exists for cases in the partial integration/late-stage effort category to prolong the policy-making process and be viewed by policy designers as a bureaucratic impediment. This occurred at the outset of the Czech Republic SEA, before the policy proponent eventually came to appreciate the value of SEA.

Integration and Concerns about Watering Down Attention to the Environment

In their book summarizing international experience with SEA, Dalal-Clayton and Sadler (2005) highlight the importance of integration. In general, they write, "an SEA process should . . . be integrated with parallel analyses of economic and social dimensions and issues, and with other planning and assessment instruments and processes" (2005: 15). Dalal-Clayton and Sadler distinguish two forms of integration embedded in the previous statement. They use the term *horizontal integration* to refer to the "bringing together of different types of impacts—environmental, economic and social—into a single overall assessment, at one or more stages in the planning cycle . . ." (2005: 369). This is different from *process integration*, which involves "integrating assessment findings into decision-making at different stages in the planning cycle" (2005: 369).[21]

In the context of table 2.2, the term *integration* is used in the sense of process integration, but both process and horizontal integration are important. As the case studies indicate, careful attention to process integration is essential if the work of SEA specialists is to be considered fully and efficiently in policy making.[22] Horizontal integration is also important, because without carefully examining linkages between environmental, social, and economic processes, it is difficult to identify environmental effects that may be indirect consequences of economic and social changes induced by a policy.

Smutney, Dusik, and Kosikova (2005: 44) describe the significance of process integration based on their experience in the Czech Republic. "Establishing good communication among the parties involved and especially between the proponent and the SEA team is a key condition for an effective process," they write. "This should begin early, with a clear and common understanding of the process and agreement on its objectives and outputs."

This emphasis on gaining a common understanding of the role of SEA in the policy-formulation process is a characteristic of the work of Jiri Dusik and his colleagues at the Regional Environmental Center for Central and Eastern Europe, headquartered in Szentendre, Hungary.[23] For a policy-level SEA, their approach involves learning the details of the policy design process (that is, which issues are to be examined, why, and in what sequence) before designing the SEA. Dusik and his colleagues ask policy makers whether the normal course of policy design includes specific environmental investigations or consultations with environmental authorities or the public. Using this knowledge, an SEA team can take advantage of opportunities to build on assessments and consultations that policy makers would undertake whether or not an SEA was to be conducted. Based on his experience with more than 15 SEAs conducted in this way, Dusik finds that such preliminary discussions with planners and policy makers often lead to incremental improvements in strategic planning and policy making.

Effective process integration also involves building relationships and gaining support for the SEA work among policy designers. The significance of communication and relationship building between SEA experts and policy designers is demonstrated in the Czech Republic case study, which also illustrates how positive experience with SEA can foster organizational learning. Based in part on what it learned through that SEA, the Czech Ministry for Regional Development took the initiative to begin an SEA process at the same time it started revising its tourism policy in 2006.

Although a strong case can be made for careful attention to both process integration and horizontal integration, some experts have qualms about integration. Dalal-Clayton and Sadler (2005: 370) raise the central issue: "Many people are concerned that the environmental dimension will be watered down in moving from SEA toward convergence with other appraisal and planning processes."

Other misgivings are illustrated by criticisms of the complete integration approach used in the SEA for the Slovak Republic's Energy Policy 2000. According to ERM (2004), the assessment departed from good practice, because there was no clear, structured assessment process; SEA results were not documented; and trade-offs were not analyzed. Lee (2006: 63) notes that "it is important [that an SEA process] preserves its own independence and integrity." The integrity of an SEA process can be maintained by requiring SEA documentation and by "independent auditing of both [the] assessment process and assessment findings (for example, by stakeholder and peer review)" (Lee 2006: 63).

SEA experts concerned with the watering down of attention given to the environment suggest that an SEA conducted independently and in parallel with policy design may be preferred, because an independent SEA privileges environmental factors. Connor and Dovers (2004: 168) wonder "whether an independent SEA through its explicit championing of the environment, may offer a more effective political strategy than IA [integrated assessment] or SA [sustainability assessment], when the latter may serve to submerge environmental (and perhaps social) considerations within the process dominated by more powerful economic ones."[24]

The position suggested by Connor and Dovers (2004) presumes that economic concerns will dominate concerns over environmental and social factors. This dominance may or may not be present. For example, economic concerns do not appear to have dominated concerns about the environment in the SEA for Energy Policy 2000 in the Slovak Republic. Moreover, some SEA experts believe that full integration of SEA into policy design provides an opportunity to bring environmentalists together with economists and social assessment experts, thereby reducing barriers across disciplines and enhancing overall policy coherence.[25]

The SEA procedure for trade policy negotiations in Canada provides an innovative way of maintaining the advantages of complete integration while avoiding any dilution of attention given to environmental factors.[26] Environmental assessment of trade policy in Canada is conducted using an approach that combines the virtues of complete integration reflected in the Slovak Republic case study and the partial integration/simultaneous effort illustrated by the Argentine case. In the context of the new SEA process for Canadian trade policy, the SEA team conducts its work in parallel with the work of trade policy negotiators, thereby privileging attention to the environment. A member of the SEA team is also a member of the negotiating team, facilitating communication between the SEA team and the trade policy negotiators.

Assessments of SEA Requirements for Policy Proposals in Canada and the Netherlands

This section analyzes policy-level SEA systems in Canada and the Netherlands, extracting lessons relevant to the practice of conducting SEAs for cabinet-level proposals in these countries.[27] Canada and the Netherlands were selected because their policy-level SEA requirements have been in place for several years and are well documented. In addition, the effectiveness of each of these SEA programs has been subject to an independent external evaluation.

Studies of the effectiveness of the national requirements for environmental appraisals of proposed policies in Canada and the Netherlands involve evaluations of a specific type of SEA application, one that is different from SEAs intended to affect the original formulation of policies. In this type of SEA, an attempt is made to have a cabinet (in Canada) or a council of ministers (in the Netherlands) consider the environment in making decisions on proposals set before them. Each instance in which this is done involves an assessment, albeit one that is typically much less elaborate and complete than the assessments conducted in the case studies examined above. The evaluations discussed below examined agency performance and made comments about how things had been working and ways they could be improved.

Audit of Canadian Cabinet Directive on SEA (2004)

In 1990 the Canadian cabinet issued a directive requiring federal departments to conduct SEAs before submitting proposed policies, plans, and programs to an individual minister or the cabinet for approval. This requirement applies only if implementation of the proposal may result in significant environmental effects.

The Commissioner of the Environment and Sustainable Development in the Office of the Auditor General of Canada assessed the performance of federal departments in implementing the 1990 directive (as amended). A 1998 assessment supported findings of an earlier study by the Canadian Environmental Assessment Agency that "most departments had not developed guidelines or directives on the Environmental Assessment of policies or programs" (Commissioner of the Environment and Sustainable Development 1998:17). It also found that SEAs were conducted without consulting experts across departments or even within the departments conducting the SEAs. Moreover, some senior departmental officials responsible for preparing cabinet documents "either were not aware of the existence of the . . . directive or did not know how it was being implemented" (1998: 18). A follow-up report in 2000 noted that departments were not "making sufficient progress to fully correct deficiencies" (Commissioner of the Environment and Sustainable Development 2000).

In 2004 the commissioner conducted another assessment, focusing on performance from 2000 to 2002. This assessment found that, in general, "departments and agencies do not know how the strategic environmental assessments they have done have affected the decisions made, and, in turn, what the ultimate impacts on the environment are" (Commissioner of the Environment and Sustainable Development 2004: 22). Moreover, audit results "suggest that most departments have not made serious efforts to apply the directive" (28).[28]

The 2004 audit closes by asking why, more than a decade after the cabinet directive was first issued, a performance gap remains. According to the report, one important reason is the lack of commitment by senior management. The report noted a correlation between relatively strong performance and high commitment by top-level managers. In departments in which senior managers lacked this commitment, the importance of the directive to the department was not communicated to the organization, expected outcomes were not articulated clearly, and sufficient resources and staff for implementation were not made available. The audit lauded performance by the Department of Foreign Affairs and International

Trade, which had a Web site that included an executive message establishing ministerial commitment to SEA, as well as information on the benefits of SEA and instructive case studies. Another factor impeding progress was the absence of a central oversight authority. In response, the report of the Commissioner of the Environment and Sustainable Development (2004) recommended that the Privy Council Office ensure that responsibilities and authorities are assigned for central monitoring and compliance with the cabinet directive on an ongoing basis. Rather than embracing this recommendation, the Privy Council Office argued that SEAs were self-assessments and that quality control could be ensured by existing mechanisms for intersectoral coordination.

The report of the Commissioner of the Environment and Sustainable Development (2004) also highlighted a systemic problem based on the lack of congruence between SEA as a "one-time-only" exercise and the existence of multiple decision-making points in the policy-making process. Thus, for example, some SEAs were conducted when a policy was first proposed and only at a general level. No follow-up was done when, much later in the process, details had been added to flesh out (and sometimes change considerably) the original policy statement. In contrast, other SEAs were conducted late in the policy-making process, just before submission for funding approval. In these cases, it was generally too late for the SEA process to inform policy design.

This timing dilemma is linked to the practice of treating SEAs as "a separate, isolated track, or 'silo,' which is not integrated with other analyses" (Commissioner of the Environment and Sustainable Development 2004: 20). This practice limits the degree to which SEA results are integrated into decision making and may result in missed opportunities for integrating environmental studies with social and economic analyses.

SEA specialists concerned that fully integrated assessments may dilute consideration given to environment should consider the alternative: if integration of SEA begins late in the policy-design process, assessment results may be discounted heavily. The 2004 audit reinforces the argument that more attention should be given to fostering both process and horizontal integration. Process integration brings SEAs more fully into policy design processes, instead of viewing them, as is currently done under the Canadian cabinet directive, as one-time exercises to assess a particular policy alternative. Horizontal integration considers environmental, social, and economic assessments together, so that indirect environmental effects can be accounted for and decision makers can be presented with more-complete assessment results.

Evaluation of the Netherlands' E-Test

Beginning in the 1980s, the government of the Netherlands experimented with requirements to accompany legislative proposals with a statement of unintended side effects.[29] In 1994 this effort was expanded in the form of an "Environmental Test" (E-Test)—a requirement to assess the environmental impacts of proposed legislation. E-Test results are attached to draft legislation submitted for consideration by the Council of Ministers.

Many Dutch laws are drafted by government departments using an informal process based on trust and cooperation among civil servants. This process is carried out without requirements for public participation and does not involve external review.

Those conducting the E-Test were asked to determine the consequences of proposed legislation on energy consumption and mobility; use of renewable and nonrenewable resources; waste and emissions into the air, soil, and water; and use of open space. E-Test results were reviewed by the Joint Support Center for Draft Legislation, which served as both a review body and a source of assistance (by providing information and environmental data). In cooperation with the Ministry of Justice, the center could oppose the submission of proposed legislation to the Council of Ministers if the E-Test did not produce information needed by the council to make an informed decision.

After the first five years of experience with the E-Test, the Ministry of Housing, Spatial Planning and the Environment hired consultants to evaluate the test's effectiveness. After reviewing documents, interviewing staff, and conducting case studies, the consultants concluded that the E-Test was carried out so late in the process of drafting a legislative proposal that it had little influence on the quality of the legislation eventually adopted. Moreover, staffs in many departments knew little about the E-Test. In general, the experience with the E-test through 2001 showed that the test produced information that "played only a limited role in policy-making and contributed little to the improvement of draft laws and regulations" (van Dreumel 2005: 73).

In response to the 2001 evaluation, the government replaced the E-Test with a two-part process. The first part of the process is a "quick scan" screening step used in the early stages of policy formulation that yields both a decision on whether additional study is needed and the TOR for the additional study. The additional study, an "environmental appraisal," is conducted by the department taking the lead in proposing the legislation.

The second part of the process is the granting of oversight responsibilities to the Ministry of Justice. After an appraisal is conducted, the Ministry of Justice prepares a "legislation report" indicating whether the quality of information yielded by the appraisal is appropriate for the draft legislation being put forward. If the Ministry of Justice finds that the information is not of sufficiently high quality, it consults the ministry proposing the legislation. Following consultation, if there is no agreement regarding the completeness of the information, this is noted in the legislation report and the "the responsible ministry includes this report in the documents for the Council of Ministers" (van Dreumel 2005: 75).

Lessons Learned

Compliance with policy-level SEA requirements in both Canada and the Netherlands can be described as pro forma. Where solid progress has been made, it has resulted partly in response to evaluations by external bodies and the commitment of senior officials in particular agencies.

Self-assessment is a central feature of the environmental assessments of policy proposals in both countries. But why would departments spend time and money on such self-assessments? Part of the explanation for the slow progress in both Canada and the Netherlands is the absence of strong incentives for departments to devote resources to policy-level SEAs.

What can be done to improve things? One approach, which has been discussed in both countries, requires creating an oversight body to control the quality of SEA results. In Canada the Commissioner of the Environment and Sustainable Development (2004) recommended that the Privy Council assign oversight authorities to a centralized body, but this recommendation was not embraced. In the Netherlands the Ministry of Justice, which has cross-sectoral responsibilities, has responsibility for commenting on the quality of information generated during the environmental assessment. The ministry's oversight powers are modest: it comments on whether the quality of information yielded by the environmental assessment is appropriate for the draft legislation being proposed and alerts the Council of Ministers when it finds an environmental appraisal inadequate. Because this oversight requirement was established only recently, it remains to be seen whether it will bear fruit.

A different strategy, which could be pursued simultaneously with the creation of an oversight organization, involves raising awareness by agencies of ways in which SEAs can be used to attain agency goals. According to the 2004 audit conducted in Canada, the commitment to SEA of high-level officials within federal departments was correlated with whether SEAs were taken seriously. If agency leaders believe SEAs can provide information that helps them meet their goals, the chances of having meaningful assessments conducted are likely to be enhanced.[30] This was clearly demonstrated by actions taken at the Department of Foreign Affairs and International Trade and several other agencies in Canada.

The evaluations in Canada and the Netherlands highlight a systemic problem with treating SEA as an exercise involving the analysis of environmental effects of a specific policy proposal. Assessment of the Canadian cabinet directive emphasized the limitations of "one-time-only" assessments in the context of policy making, which involves multiple decision-making points. The evaluation of the E-Test in the Netherlands noted that the test was often conducted too late in the process to have much influence on legislation. These systemic difficulties suggest a need for beginning environmental assessment earlier and either fully integrating SEA into policy design or employing partial integration with a parallel SEA effort that continues for the duration of the policy-design process.

Conclusions

What does it take for SEA to fulfill the goal of integrating environmental considerations into policy designs? Some aspects of the answer to this question are self-evident. SEAs must be carried out by trained staffs using appropriate methods and supported by adequate resources. This chapter does not focus on these matters. Instead, it establishes the overarching importance of integrating SEA into policy-design processes and creating incentives for policy proponents to take SEA seriously.

Before presenting conclusions from this analysis, a caveat is in order. No inquiry based on the modest set of case studies and evaluations considered here can yield anything more than hypotheses requiring further investigation.

Process integration challenges are closely tied to the incentives of policy proponents to conduct SEAs and employ assessment results. Where incentives are weak, pro forma compliance (or even noncompliance) with formal SEA requirements may be widespread. In contrast, policy proponents motivated to embrace SEA because they see value in doing so are likely to give careful attention to process integration issues.

Fostering Interaction between SEA Teams and Policy Designers

How can SEA processes be structured to allow SEA teams to interact with policy designers in ways that facilitate consideration of the environment in policy making? Each of the case study settings in which SEA had the greatest influence on policy making—Argentina, the Czech Republic, and the Slovak Republic—involved intensive interaction between the SEA team and policy designers. In Argentina and the Slovak Republic, SEA was initiated early in the policy formulation process, and SEA specialists interacted frequently with policy designers.[31] In the Czech Republic the SEA started late; although it nevertheless influenced policy making (because of the skill and effort the SEA team employed in fostering open communications and building trust with the policy designers), there were notable delays in the policy-making process.

The evaluations of environmental assessment requirements for cabinet-level proposals in Canada and the Netherlands reinforce the overriding importance of process integration. In both cases compliance with requirements was pro forma, and there was little apparent influence on outcomes as a result of most assessments. In the Canadian case, the evaluation emphasized the limitations of conducting one-time-only assessments of policies at any particular stage in the policy-formulation process. In the Netherlands the lack of influence of environmental assessments on proposed legislation was attributed to the fact that the assessments were submitted too late (just before submission of legislative proposals to the Council of Ministers). Revisions of the process addressed this weakness by introducing a quick scan, conducted early in the policy-making process so that a more complete environmental appraisal could be conducted when warranted.

Some SEA experts argue that integrated assessments may dilute the attention given to environmental factors. In the cases reviewed here, there was no evidence of such dilution. It is not clear that integration (in the sense of both process integration and horizontal integration) need involve dilution, because innovative processes can be created to offset any such danger. This is demonstrated by the new SEA approach developed by Foreign Affairs Canada for conducting environmental assessments for trade policies.

Creating Incentives for Compliance

What steps can be taken to provide policy designers with incentives to use environmental assessment results in policy making? The importance of this question is established by noting what has happened in circumstances in which incentives were weak (in the sense that failing to comply with SEA requirements was not penalized). In Canada the absence of penalties for noncompliance is viewed as an important reason for the pro forma compliance of many agencies to the requirement for environmental assessments of cabinet-level proposals.

One approach to creating incentives for complying with policy-level SEA requirements involves imposing penalties for failure to comply. A standard way to do this is to give a cross-sectoral agency responsibility and tools for ensuring compliance.

Another way of creating incentives involves controlling access to resources, an approach demonstrated by the Argentine case study. The government of Argentina could not have obtained the loan it sought if it had not conducted the environmental assessment required by the World Bank's internal directives.

These incentives are external to the policy proponent: oversight is exercised by another agency and control over resources is exerted by a funder. Incentives can also be internal. Based on its experience with SEA in creating a national tourism policy, the Czech Republic's Ministry for Regional Development went from being a reluctant participant to a proactive adopter of the new environmental assessment process at the very start of its recent effort to revise its tourism policy. A similar reversal in attitude toward SEA occurred in Argentina, where the Ministry of Economy, which initially had reservations about the need for SEA, became an SEA supporter after seeing how the assessment could help meet the ministry's objectives in providing drinking water supplies. Similar outcomes are evident in Canada, where leaders of agencies that have embraced environmental assessment requirements for cabinet-level proposals view SEA as an instrument for goal attainment. For example, the Canadian Ministry of Foreign Affairs used its early experience with SEAs for trade policies to create an innovative SEA process to help meet its sustainable development goals.

Fostering Organizational Learning

What can be done to foster organizational learning? Carefully tailored case studies used to inform agency leaders may provide an answer.

Policy proponents might take SEA seriously if they learned of case studies demonstrating the value added by SEAs conducted in circumstances similar to their own. The challenge is in identifying case study information relevant to agency-specific contexts and in finding venues at which policy proponents can be informed about such case studies. This information transfer challenge might be handled at the national and subnational levels of government, where it is possible to provide case study material and training programs tailored to circumstances faced by individual agencies.

Gaining High-Level Commitment

Gaining the commitment of high-level agency officials is a critical step in implementing a program requiring policy-level SEAs. As experience in Canada demonstrates, commitment by agency heads signals to lower-level staff that SEA is to be taken seriously. In these circumstances, agency resources can be reallocated and SEA processes crafted to reflect the realities of day-to-day agency operations.

Notes

1   For an introduction to the burgeoning literature on SEA, see Dalal-Clayton and Sadler (2005). Jones and others (2005) document the extensive use of SEA in land use planning.

2   Examples of countries with experience with policy-level SEAs dating to the 1990s include Canada, the Czech Republic, Denmark, Finland, and the Slovak Republic.

3   The definition of *policies* suggested by Dalal-Clayton and Sadler (2005: 18)—"broad statements of intent that reflect and focus the political agenda of a government"—is adopted here. Policies include "laws, regulations, strategies and other official/formal government processes that have an implication on subsequent actions" (ERM 2004: 2).

4   This chapter summarizes an unpublished report originally prepared as a contribution to World Bank (2005). The report was prepared in response to contract terms of reference that singled out the case studies in ERM (2004) and the SEA for Argentina discussed below as well as the evaluations in Canada and the Netherlands considered here.

5   For insights into the challenges involved in conducting ex post studies to determine the influence of SEAs on environmental change, see Partidário and Fischer (2004) and Partidário and Arts (2005).

6   This section is based on Sánchez-Triana and Enriquez (2005).

7   Under current Canadian practice, environmental reviews for treaties begin before or, at the latest, at the beginning of negotiations (J. Shuttleworth, Foreign Affairs Canada, personal communication, June 17, 2005).

8 This paragraph is based on Hazell and Benevides (2000) and ERM (2004).

9 In 2003 DFAIT was divided into two departments, Foreign Affairs Canada and International Trade Canada. The SEA procedure is detailed in DFAIT (2002).

10 General information on SEA in the Czech Republic in this paragraph and the two that follow is from Machac, Rimmel, and Zenaty (2000), and from personal communications with J. Dusik, Regional Environmental Center for Central and Eastern Europe, June 1 and 18, 2005.

11 The Ministry for Regional Development's feared delay in the policy-making process did occur, but it resulted, in part, from the ministry's initial omission of required SEAs.

12 Information in the remainder of this subsection is from J. Dusik, SEA consultant (personal communications, May 25 and June 18, 2005).

13 Another explanatory factor, which Dusik considers minor, is that the potential for finding mutually supportive linkages between tourism development and environmental protection makes it possible to identify actions that advance tourism while protecting the environment.

14 These changes were accepted in less than an hour, because targets for the protection of environment had already been agreed on in the SOP.

15 At the end of the SEA process, the SEA consultants and the two ministries agreed that the ministries would later finalize the monitoring system jointly and clarify institutional responsibilities for monitoring. The rapid staff turnover that followed had not been anticipated.

16 Marshall and Fischer (2005) give another example of organizational learning in the context of SEA. They show how ScottishPower used SEA to "not only address the environmental impact of future investment programmes but . . . also to enhance the environmental governance and stewardship of established corporate decision-making frameworks" (687–88). They conclude that, for purposes of evaluating "environmental parameters on equal footing with economic and technical parameters. . . . [SEA] is a suitable instrument that is able to strengthen and improve corporate decision-making procedures" (688).

17 This section is based on ERM (2004).

18 The government had experience conducting SEAs for energy policy. That experience influenced the approach followed in 1999 when a new government started revising the energy policy.

19 Many others have highlighted the significance of SEA process integration (see, for example, João 2005 and Partidário 2005).

20 The effects of SEAs on monitoring are difficult to discern for the NAFTA and Argentine cases.

21 Dalal-Clayton and Sadler (2005: 369) cite a third form of integration, which they call *vertical integration*: the integration of assessments "undertaken at different stages in the policy, planning and project cycle (tiering)." Scrase and Sheate (2002) go farther, distinguishing 14 meanings of *integration* in the context of environmental governance.

22 Having an influence on policy design is a key reason for conducting a policy-level SEA.

23 In elaborating on practical issues linked to careful integration of SEA into planning and policy making, Dusik cautions that flexibility is required, because planners and policy makers may change their ideas abruptly and new SEA studies may be needed in response to these changes. SEA budgets and TORs need to reflect these possible outcomes (personal communication January 19, 2006).

24 *Integrated assessment* is often used to refer to horizontal integration. *Sustainability assessment* is "an integrated assessment that is carried out within an explicit framework of *sustainability*

objectives and criteria" Sadler (2005: 3, emphasis added). For more on sustainability assessment, see Gibson (2005).

25 This view was expressed by Jaye Shuttleworth (Foreign Affairs Canada) and Colin Kirkpatrick (University of Manchester) in response to questions at a forum on "Ethics and Quality in Trade Impact Assessment" at the June 2005 meeting of the International Association for Impact Assessment, in Cambridge, MA. The journal *Integrated Assessment* contains numerous examples documenting the growing experience with integrated assessment.

26 For more information on the Canadian procedures for SEA in the context of trade policy negotiations, see DFAIT (2002).

27 This assessment was carried out in 2005, relying on the information available at the time. It does not take into account any changes to these SEA systems in those countries made after June 2005.

28 The 2004 audit also contained some positive findings and cited good practices. Some departments were lauded for accountability structures, guidance (both in documents and on Intranets), and SEA screening and tracking systems.

29 Except where noted, this section is based on Verheem (2004).

30 Marshall and Fischer (2005) examine how companies can use SEAs to meet their goals. Their discussion points to specific benefits from the voluntary application of SEA, including time and cost savings reaped by establishing a strategic decision-making framework that consolidates existing procedures and facilitates participation and consultation. They also note increased "levels of public understanding and acceptance of the plans and programs. . . . Furthermore, SEA may assist in reducing negative environmental impacts or delays in corporate objectives that are ultimately costly for private companies" (676).

31 The Slovak SEA involved complete integration; the Argentine case involved partial integration/simultaneous effort.

References

Commissioner of the Environment and Sustainable Development. 1998. "Environmental Assessment: A Critical Tool for Sustainable Development." In *Report of the Commissioner of the Environment and Sustainable Development to the House of Commons*. Office of the Auditor General of Canada, Ottawa.

———. 2000. "Follow-Up of Previous Audits: More Action Needed." In *Report of the Commissioner of the Environment and Sustainable Development to the House of Commons*. Office of the Auditor General of Canada, Ottawa.

———. 2004. "Assessing the Environmental Impacts of Policies, Plans, and Programs." In *Report of the Commissioner of the Environment and Sustainable Development to the House of Commons*. Office of the Auditor General of Canada, Ottawa.

Connor, R., and S. Dovers. 2004. *Institutional Change for Sustainable Development*. Cheltenham, United Kingdom: Edward Elgar.

CSIR (Council for Scientific and Industrial Research). 2000. *Guideline Document, Strategic Environmental Assessment in South Africa*. Report prepared in cooperation with the Department of Environment and Tourism, Pretoria.

Dalal-Clayton, B., and B. Sadler. 2005. *Strategic Environmental Assessment: A Sourcebook and Reference Guide to International Experience*. London: Earthscan.

DFAIT (Department of Foreign Affairs and International Trade). 2002. *Handbook for Conducting Environmental Assessments of Trade Negotiations*. Ottawa.

Dusik, J., and S. Kosikova. 2004. "Designing Effective SEA Systems: Lessons from REC Pilot SEA Projects." In *Strategic Environmental Assessment Making a Difference*, publication 2003/012, European Environmental Bureau, Brussels.

ERM (Environmental Resources Management). 2004. "Case Studies of Policy SEAs." Report prepared for the World Bank, Environment Department, Washington, DC.

Gibson, R. B. 2005. *Sustainability Assessment: Criteria, Processes and Applications*. London: Earthscan.

Hazell, S., and H. Benevides. 2000. "Toward a Legal Framework for SEA in Canada." In *Perspectives on Strategic Environmental Assessment*, eds. R. Partidário and R. Clark, 47–68. Boca Raton, FL: Lewis Publishers.

João, E. 2005. "Key Principles of SEA." In *Implementing Strategic Environmental Assessment*, eds. M. Schmidt, E. João, and E. Albrecht, 3–14. Berlin: Springer.

Jones, Carys, Mark Baker, Jeremy Carter, Stephen Jay, Michael Short, and Christopher Wood, eds. 2005. *Strategic Environmental Assessment in Land Use Planning*. London: Earthscan.

Lee, N. 2006. "Bridging the Gap between Theory and Practice in Integrated Assessment." *Environmental Impact Assessment Review* 26 (1) 57–78.

Machac, M. D., V. Rimmel, and L. Zenaty. 2000. "SEA in the Czech Republic." In *Perspectives on Strategic Environmental Assessment*, eds. M. R. Partidário and R. Clark, 81–89. Boca Raton, FL: Lewis Publishers.

Marshall, R., and T. Fischer. 2005. "Best Practice Use of SEA: Industry, Energy and Sustainable Development." In *Implementing Strategic Environmental Assessment*, eds. M. Schmidt, E. João, and E. Albrecht, 673–72. Berlin: Springer.

Partidário, M. R. 2005. "Capacity-Building and SEA." In *Implementing Strategic Environmental Assessment*, eds. M. Schmidt, E. João, and E. Albrecht, 649–63. Berlin: Springer.

Partidário, M. R., and J. Arts. 2005. "Exploring the Concept of Strategic Environmental Assessment Follow-Up." *Impact Assessment and Policy Appraisal* 23 (3): 246–57.

Partidário, M. R., and T. B. Fischer. 2004. "Follow-Up in Current SEA Understanding." In *EIA and SEA Follow–Up*, eds. A. Morrison-Saunders and J. Arts, 97–117. London: Earthscan.

Rossouw, N., and K. Wiseman. 2004. "Learning from the Implementation of Environmental Public Policy Instruments after the First Ten Years of Democracy in South Africa." *Impact Assessment and Project Appraisal* 22 (2): 131–40.

Sadler, B., ed. 2005. *Strategic Environmental Assessment at the Policy Level: Recent Progress, Current Status and Future Prospects*. Szentendre, Hungary: Regional Environmental Center for Central and Eastern Europe.

Sánchez-Triana, E., and S. Enriquez. 2005. "Using Strategic Environmental Assessments for Environmental Mainstreaming in the Water and Sanitation Sector: The Cases of Argentina and Colombia." Environmental and Socially Sustainable Development, Latin America and Caribbean Region, World Bank, Washington, DC.

———. 2007. "Using Policy-Based Strategic Environmental Assessments in Water Supply and Sanitation Sector Reforms: The Cases of Argentina and Colombia." *Impact Assessment and Project Appraisal*. 25 (3): 175–87.

Scrase, J. I., and W. R. Sheate. 2002. "Integration and Integrated Approaches to Assessment: What Do They Mean for the Environment?" *Journal of Environmental Policy & Planning* 4 (4): 275–94.

Shuttleworth, J. 2005. "Building on Experiences and Addressing Challenges." Paper presented at the "Theme Forum on Ethics and Quality in Trade Impact Assessment," 25th Annual Conference of the International Association for Impact Assessment, Cambridge, MA, May 31–June 3.

Smutney, M., J. Dusik, and S. Kosikova. 2005. "SEA of Development Concepts in the Czech Republic." In *Strategic Environmental Assessment at the Policy Level: Recent Progress, Current Status and Future Prospects,* ed. B. Sadler, 36–45. Szentendre, Hungary: Regional Environmental Center for Central and Eastern Europe.

van Dreumel, M. 2005. "Netherlands' Experience with the Environmental Test." In *Strategic Environmental Assessment at the Policy Level: Recent Progress, Current Status and Future Prospects,* ed. B. Sadler 69–75. Szentendre, Hungary: Regional Environmental Center for Central and Eastern Europe.

Verheem, Rob. 2004. "Evaluation of the Dutch E-Test." Commission for Environmental Impact, Utrecht, the Netherlands.

Wiseman, K. 2000. "Environmental Assessment and Planning in South Africa: The SEA Connection." In *Perspectives on Strategic Environmental Assessment,* eds. M. R. Partidário and R. Clark, 155–66. Boca Raton, FL: Lewis Publishers.

World Bank. 2005. *Integrating Environmental Considerations in Policy Formulation: Lessons from Policy Based SEA Experience.* Report 32783, Environment Department, Washington, DC.

CHAPTER 3

The Continuous Process of Policy Formation

Martha S. Feldman and Anne M. Khademian

THIS CHAPTER EXAMINES two ways of conceptualizing policy formation.[1] The first is as a rational decision-making process, exemplified by current use of Strategic Environmental Assessment (SEA). SEA is a tool for integrating environmental considerations in the formation of programs, plans, and policies. It is built upon an understanding of policy formation as linear stages of rational decision making focused on defining policy problems and identifying and selecting solutions. Drawing on the example of SEA, this chapter introduces three assumptions that are key to understanding models of rational decision making and shows how the uncertainty and ambiguity of policy formation make these assumptions hard to meet.

Martha S. Feldman is the Johnson Chair for Civic Governance and Public Management in the Department of Planning, Policy, and Design at the University of California–Irvine. Anne M. Khademian is associate professor at the Center for Public Administration and Policy School for Public and International Affairs, at Virginia Tech, in Alexandria, VA. The authors thank Andrea Ballestero for sharing her research on reform of the water sector in Ceará, Northeast Brazil, and for her careful review of this chapter to integrate that research. Kathy Quick read and provided very helpful comments on the chapter. Cressa Paz helped prepare the references. The authors are grateful for the comments of two anonymous reviewers on an earlier version of this chapter. Their thanks also go to Kulsum Ahmed and Ernesto Sánchez-Triana for the opportunity to contribute to this volume and for the guidance, suggestions, and helpful feedback they provided.

Alternatively, policy making can be conceptualized as a continuous process in which an action taken at one point is not an end but a means; each action helps reveal the next set of policy questions and possibilities. For example, the action of building taller smokestacks reduced the amount of air and ground pollution for communities near the smokestacks, but it revealed the next set of questions, related to the complex problem of acid rain affecting communities far from the initial pollution source.

The model of policy formation as a continuous process combines two models that have been developed elsewhere. One of these models, "adaptive management," is necessary because action must be taken even when there is considerable uncertainty about its effects (Holling 1978; Walters and Holling 1990). Adaptive management is a way of taking action designed to decrease risk and increase opportunities for learning. The other model, "inclusive management," describes how people with different orientations and ways of understanding and valuing policy problems and solutions can work together to propose and take action (Feldman and Khademian 2000, 2007).

Central to understanding policy formation as a continuous process is an appreciation of policy processes as fundamentally social. Policy problems involve people in addition to material objects. They are complex and require multiple sources of knowledge and information to address. They require getting people to move from one set of policy actions to another. These fundamentally social aspects of policy formation point to the ongoing creation of a *community of participation* in any policy process and the importance of the qualities of any community of participation for taking action in the context of ambiguity.

Using the case study of a small community in Northeastern Brazil (Ballestero 2004, 2006), the chapter shows how the continuous process incorporates both adaptive and inclusive management and how the combination enables policy formation to encompass more viewpoints, including those that have been marginal in other processes. The chapter is organized as follows. The next section explores two models of policy formation, the rational decision-making model and the "garbage can" model of decision making. The second section examines policy formation as a continuous process in which both uncertainty and ambiguity must be managed. The last section draws some conclusions.

Policy Formation as Decision Making

SEA is a tool for injecting environmental concerns into the discussion of policies, plans, and programs that have significant environmental implications. Rather than assessing concerns at the project level, where critical decisions of design and location have already been made, SEA seeks to influence decisions "upstream," so that environmental considerations are reflected in choices for alternative projects (World Bank 2002).

SEA is often described as a systematic process. It is built on linear assumptions about the development of public policy in progressive stages and the rational capacities of individuals and organizations making the decisions. It involves the following steps:

- Identifying key environmental impacts through a screening and scoping exercise
- Assessing all concerns in a report that assembles information, considers alternatives, analyzes the potential impacts associated with all alternatives, and identifies measures to mitigate them
- Making decisions and implementing recommendations following discussions with stakeholders
- Monitoring and discussing results with stakeholders.

In practice, implementing this formal process has been difficult, and the incorporation of SEA findings into decisions made at the program and policy levels has been limited (Dalal-Clayton and Sadler 2005). Efforts to address the limitations of SEA as a means of influencing policy upstream are ongoing (Stockholm Environment Institute 2003). This is particularly important as development organizations try to identify the synergies between poverty alleviation, sustained economic growth, and environmental protection and to assess trade-offs between these objectives.

The limitations of SEA are intrinsic to the assumptions about the dynamics of policy formation embedded in the tool. For this reason, this chapter proposes alternative models of public policy formation rather than adjustments to SEA.

Models of decision making follow a continuum in the assumptions they make about the degrees of certainty and ambiguity confronting decision makers (figure 3.1). At one end of the continuum are assumptions that decision makers are unitary actors with clear preferences and goals, that they have perfect information, and that problems are clearly defined. Across the continuum, these assumptions give way to uncertainty and ambiguity. Of course, any model is a

FIGURE 3.1
Spectrum of Decision-Making Models

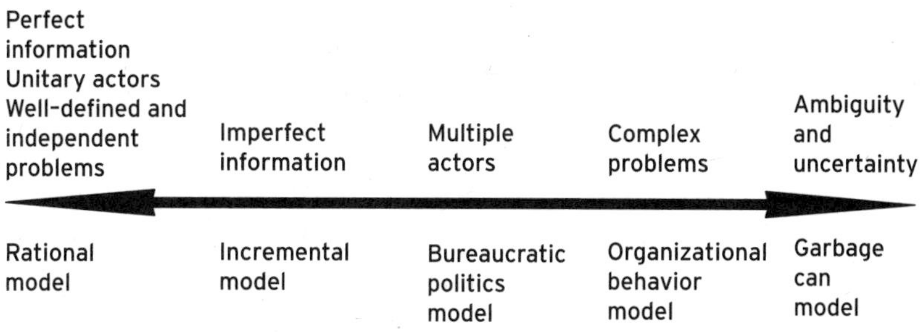

Perfect information Unitary actors Well-defined and independent problems	Imperfect information	Multiple actors	Complex problems	Ambiguity and uncertainty
Rational model	Incremental model	Bureaucratic politics model	Organizational behavior model	Garbage can model

Source: Authors.

simplification. For this reason, the focus is not on the particular ways in which a model works in practice—which will always deviate from the abstract description—but on the basic assumptions. If the basic assumptions are fundamentally disconnected from the context, even at the most abstract level, the model is not an appropriate base for action.

The Rational Decision-Making Model

SEA is very similar to the rational decision-making model (table 3.1). Its sequential process makes sense in the abstract. It is in the practice of decision making or policy formation that difficulties are encountered. Many of the problems derive from the assumptions implicit in rational models. These assumptions fall into three categories: assumptions about information, assumptions about unitary actors, and assumptions about the nature of the problems being addressed.

The second step of the rational decision-making model shown in table 3.1 is not explicitly included in SEA, but it is implicit in the identification of key environmental impacts. The other steps of the rational model are all reflected in the SEA model.

Assumption about information. The rational decision-making model assumes that a decision maker has complete information. For many reasons, this assumption does not hold in most decision-making and policy-making processes. The fundamental problem with this assumption is that people are limited information processors (Simon 1957a, 1957b; March and Simon 1958). The classical rational model requires an amount of simultaneous information processing that is simply beyond the capacity of human beings (Steinbruner 1974). The demand can be met if the boundaries of the problem are drawn very narrowly

TABLE 3.1
Similarities between Rational Decision-Making Model and Strategic Evaluation Assessment

Rational Decision-Making Model	Strategic Evaluation Assessment
1. Identify the problem	1. Identify key environmental impacts
2. Establish preferences	2. (Implicit in identification of key environmental impacts)
3. List all options or alternatives	3. Assess and consider alternatives and measures to mitigate
4. Gather all relevant information	4. Gather information
5. Make choice that maximizes or optimizes the likelihood or efficiency of achieving preferences	5. Make choice
	6. Monitor impact of choice (repeat steps 4 and 5)

Source: Authors.

(for example, what is the best policy for X given criterion Y?), but decision making seldom takes place within such narrow confines: policy formation is primarily about coming up with the appropriate X and Y. The demand is not just to understand everything there is to know about a limited domain but to understand many different perspectives. For this reason, although computers have greatly increased the capacity to process information, they have not fundamentally altered this limitation.

Simon (1957a) and March and Simon (1958) describe this basic limitation as *bounded rationality* (rationality is bounded by cognitive limitations) and the response to it as *satisficing*. Satisficing involves choosing an option that satisfies, rather than maximizes or optimizes, the prespecified criteria. It is a change in the fifth and last step of the rational decision-making model. But this apparently minor change is actually the beginning of the end of the rational model. In this bounded rational decision-making model, all relevant information is not gathered. All alternatives are not considered. Rationality resides instead in the definition of the problem and the establishment of preferences, both of which are clearly the outcome of similarly constrained decision processes based on similarly limited informational resources.

Assumption of unitary actors. Rational-decision making models assume that policy-making entities or relevant decision makers (organizations, institutions, nation-states, networks) are unitary actors with preferences (described by utility functions) that are clear, consistent, and stable and that the decisions these actors make are consistent with those preferences (March 1978). In the study and practice of international relations, nation-states are often viewed as unitary actors that have clear, consistent preferences and that take actions reflecting those preferences (Wolfers 1962). The evidence is overwhelming, however, that policy-making entities are not unitary actors but are constituted of multiple actors with multiple and often conflicting goals (Cyert and March 1963; Allison 1971; Halperin 1974). In the case of nation-states, conflicting interests within legislatures, between the executive and legislative branches of government, between political parties, and between coalitions of interests complicate decision processes (Hug 1999).

The fact that policy is made through the interactions of multiple actors with multiple and conflicting goals affects the rational decision-making model in a number of ways. It means determining the problem and establishing preferences are likely to involve political as well as technical considerations. A coalition is generally necessary for taking either of these steps. Building coalitions involves politics, power, and conflict (Stevenson, Pearce, and Porter 1985; Flyvbjerg 1998). As a result, policies are often aimed not only at solving particular problems but also at holding together a coalition of support. The foundation of the rational process is thus built on processes that are fundamentally nonrational.

While coalitions may be built around particular goals by compromising, it is possible to proceed to action without resolving the conflicts between preference functions. A different way of building coalitions—thereby dealing with multiple and conflicting goals—involves an even more serious break with the rational model. Cyert and March (1963) note that the "sequential attention to goals" is one way of dealing with multiple goals. Logrolling is the political form of this strategy. Lindblom (1959) shows that sometimes policy makers simply ignore differences in preference functions and focus instead on policy. He illustrates this by giving examples of policies that satisfy multiple and even opposing preference functions.

Two models of policy formation represent a response to the realization that policy decisions are not made by unitary actors: the governmental (or bureaucratic) politics model and incremental decision making. In the first model, decision makers are multiple unitary actors. This type of model represents the notion that "where you stand depends on where you sit" (Allison 1971; Halperin 1974). People in an environmental agency tend to know more and care more about the effect of a policy on air or water quality, while people in agencies that deal with commerce or defense tend to know more and care more about the effect of a policy on the economy or national security. Who takes the lead on a policy and who has to sign off on a policy will make a critical difference to what information is produced (Feldman 1989) and what policies are made (Seidman 1980). People with access to the process will try to influence policy in the direction of their preferred policies. Policies are often the result of compromise among a variety of competing preferences.

In the incremental decision-making model, the legacy of past policy decisions conditions the present policy-making process. Unlike in the rational model, decision makers in the incremental model do not reconsider policy preferences and consider a broad array of possible alternatives to current policies. Rather, they focus on aspects of the policy that are considered problematic and make marginal adjustments in existing programs to resolve these problems. The difference is illustrated by efforts to engage in rational budgeting processes in which budgets are built from the ground up rather than adjusted incrementally. Proponents of incremental policy making argue that this process allows policy makers to make sensible adjustments to fine-tune programs and policies and that the alternative requires more information and more agreement than is likely to be feasible at a reasonable cost (Lindblom 1959; Wildavsky 1979). Opponents of the incremental approach claim that it institutionalizes inertia and legitimates ineffective policies (see Hochschild 1984; Gersick 1991; Weick and Quinn 1999).

Assumptions about the nature of the problem. In the rational decision-making model, problems are assumed to be well defined and independent. In fact, as problems become more complex, it is often difficult or impossible to define them clearly. How should one define policy problems that cut across efforts to preserve the environment, alleviate poverty, and support sustained economic growth? There are difficult trade-offs between preferences and alternatives, and the relevance and value of information may vary across problems or parts of problems.

Ways of coping with the complexity of problems have developed at both the micro and macro levels. At the macro level, policy makers focus on process rather than outcomes. The organizational behavior model suggests that decisions are made and policies formulated by assessing appropriate processes rather than applying conscious rationality. The way the model works has been likened to the way a thermostat works (Steinbruner 1974). An indicator triggers a response that is often worked out in advance. Events and time are often the triggers for a response. The response is often in the form of a process or routine. The *Challenger* explosion and the Cuban missile crisis, for example, set off a number of different routines that constituted the response to these events (Vaughan 1997; Allison and Zelikow 1999). Budget processes and program reviews are examples of organizational processes that tend to be organized by time.

At the micro level, individuals use intuition or rely on extensive experience in problem solving and lessons learned from successful and failed efforts (Isenberg 1984). The intuitive decision maker relies on a deep understanding of the context or setting of a problem to process information quickly in order to make a decision. Intuition, sometimes described as a "gut feeling" (Mintzberg 1994), can be used to identify problems; define them; and identify, select, and implement a solution (Isenberg 1984). Experience-based intuition, in other words, is an alternative to the extensive analysis and consideration of alternatives that would be required for rational decision making.

The Garbage Can Model of Decision Making

The garbage can model of decision making takes account of the fact that the rational decision-making model does not fit the complex world of policy making (Cohen, March, and Olsen 1972; March and Olsen 1976). It is based on the following assumptions:

- Preferences are problematic. For a variety of reasons, people do not always (or even often) know what they want. Preferences may be difficult to articulate; people may recognize their preferences only when confronted with choices. Moreover, people may not know what preferences they will have for future conditions, and different people may want different things.
- Technologies are unclear. Even when people know what they want, they may not know how to make it happen.
- Participation is fluid. The people involved in decision making change over time and in relation to different aspects of the policy.

Policy formation characterized by these assumptions is fundamentally different from rational decision making. Choice, the focal point of rational decision making, is not as central in the garbage can model. Instead, decision making is viewed as consisting of streams of problems, solutions, participants, and choice opportunities. As problems and solutions are not actually differentiated until they are connected to one another, this picture can be simplified somewhat by saying that there are streams of issues. Issues are anything that can be defined as either a problem or a solution. A dam, for instance, can be a solution to the problem of downstream flooding or upstream irrigation. It can be a problem for communities living near the current riverbed or for endangered species.

Issues and participants flow into choice opportunities. A choice opportunity occurs whenever people have a chance to connect different issues to one another. Meetings are examples of relatively formal choice opportunities. Such formal choice opportunities often have labels, such as "meetings about the budget crisis." Participants often bring issues to choice opportunities—concerns about cutting funding to social services, balancing the budget, maintaining the city's bond rating and its capacity to borrow—but other issues may be activated during the meeting when raised by a participant or when a connection is made through the discussion. Not all of the issues in the choice opportunity will be relevant to the label or seen as relevant by all participants. In a meeting about a city budget crisis, for instance, discussions of the school board election are essentially irrelevant, as the school board has a separate budget. It is nevertheless possible that this and similar issues will find their way into the discussion.

Decisions may or may not be made, and decisions that are made may or may not resolve problems. While choice opportunities provide arenas for discussing and possibly making decisions, many choice opportunities do not result in decisions, and many decisions are made outside of choice opportunities.

Decisions can be made in three ways: flight, oversight, and resolution. Of the three, resolution is the only one that bears a resemblance to the rational model. Resolution occurs when an issue that is defined as a problem and an issue that is defined as a solution are connected so that the solution resolves (or appears to resolve) the problem. Flight occurs when an issue that is making it difficult to make a choice flies from one choice opportunity to another, making it possible to make a choice. Oversight occurs when there are no issues that make it difficult to make a choice.

The garbage can model incorporates the previously described models of policy formation. Satisficing takes place in the process of resolution, which may be but is not necessarily an optimizing process. Bureaucratic politics take place in the processes of jockeying for attendance in choice opportunities and in the ways issues are defined as either problems or solutions in the various discussions that take place. Incremental decision making occurs when the choice opportunity is defined either formally or informally as adjustments to the status quo. Organizational processes, often in the form of organizational routines, influence when a choice opportunity arises as well as the repertoire of responses that are offered as problems and solutions.

The garbage can model and policy making. Kingdon (1995) builds on the garbage can model to describe the way in which agendas are set by the federal government in the United States. According to him, the process is not linear but an organized anarchy consisting of three process streams. In one stream problems are recognized, defined, and redefined. Indicators or measurements, dramatic events, and evaluations of existing programs combine with values and beliefs to bring a problem to the fore.

A second stream consists of a "policy primeval soup," the ingredients of which are ideas generated by multiple actors in multiple settings. Policy communities consisting of academics, policy analysts, scholars at think tanks, administrators, interest groups, and congressional staff who share common interests in homeland security or transportation, for example, generate ideas that they shared at professional conferences, roundtables, and public hearings; in publications; and through lectures. The policies that bubble to the top of the soup will be viewed as feasible and in line with the values and beliefs held by policy makers.

The third stream is politics. Kingdon argues that the "national mood," "pressure group campaigns," and the turnover of legislative and administrative offices define the dynamics of this stream.

Just as choice opportunities provide decision-making opportunities in the garbage can model, "windows of opportunity" provide opportunities to link the three streams and set the agenda in Kingdon's model. A "policy entrepreneur" plays the role of linkage by highlighting a problem from the problem stream, connecting it to a policy, and building on the national mood, an election, or the strength of an interest group campaign to bring the issue to the federal agenda. An advocate for election reform, for example, could view a controversial presidential election as a window of opportunity, linking the problem of ensuring fair elections with possible solutions (such as touch-screen voting machines or early voting) and building on the national mood and the interests of elected officials in winning reelection to set the agenda.

This model draws attention to the use of information as symbol, signal, and repertoire (Feldman and March 1981; Feldman 1989). Gathering and displaying information may have as much to do with the process of legitimizing the role of particular players in a policy-making process as it does with answering policy questions. Scholars have noted that the expectation that policy information will be used to solve current policy problems is unrealistic and seldom met (Weiss 1977, 1978, 1980; Lynn 1978; Lindblom and Cohen 1979; Bozeman 1986; Feldman 1989).

The continued production of information can be explained in several ways. First, attention to processing information symbolizes participants' adherence to a set of values consistent with the rational model. This adherence can legitimize the role of the participant as an appropriate decision maker (Feldman and March 1981).

Second, attention to processing information can be a signal that the participant knows how to use information in ways that will help make rational decisions. The cost of signaling decreases as the skill of the participant with information increases (Feldman and March 1981). Third, the information may form a repertoire, much like a library, that can be drawn on when needed, which is not necessarily the time frame in which it was produced (Feldman 1989). In all of these senses, information becomes another stream that flows in and out of the garbage can model. It is important to the overall outcome, but it does not have a linear relation to problem solving or policy formation.

Policy formation and path dependency. The garbage can model sheds light on the relation between policy formation and path dependence. Path dependency is a way to explain the longevity or ingrained nature of a policy. Once a policy is initiated, structures, processes, and understandings of the policy and the problems the policy addresses can take hold that define a gradually well-worn path (Wilsford 1994).

Two factors will influence whether historical legacy or novelty are prominent in policy formation at any point in time. One is attention, or the presence of an issue in a choice opportunity. As both the originators of the garbage can model and those who have used it, such as Kingdon, note, attention is critical to the process of choice. If the participants in a choice opportunity do not attend to an issue, it cannot be framed as either a solution or a problem, and it cannot be matched with a complementary part. Both history and novelty play roles in the process of attention. Some issues are attended to because they have always been associated with the choice opportunity. The state of the economy, for example, is always an issue in a national budget process, just as campaign finance has become a regular issue in a national election. Other issues gain attention in more dramatic ways. Massive flooding focuses attention on the issue of flood plain insurance. The events of September 11 focused attention on terrorism.

Events alone, however, do not determine how attention will be focused or what the outcome of such focus will be. How events are interpreted and used depends on the availability of ideas about how to move forward (*repertoires*, in Feldman's terms) as well as people in positions of power who want to move in that direction (*policy entrepreneurs*, in Kingdon's terms). One could argue that the difference between the reaction to the bombing of the Murrah Federal Building in Oklahoma City and the bombing of the World Trade Center in New York was the existence of policy entrepreneurs with plans that could be connected to the issue of terrorism from outside the United States in the case of the 9/11 attacks and the absence of both following the bombing in Oklahoma City. The political skill of policy entrepreneurs as well as serendipity—being in the right place at the right time—will influence their ability to draw attention to issues. Developing repertoires of actionable ideas requires considerable time and effort by government agencies, think tanks, and universities.

"Competency traps" are another aspect of path dependency. The development of skills, technology, and organizational routines around particular policy choices can make it difficult to move to other choices and encourage responses to novel situations that are drawn predominantly from what decision makers already know how to do. The Norwegian response to the need to regulate oil rigs in the North Sea after the discovery of oil there, for example, was to consider an oil rig "a somewhat peculiar ship" (March and Olsen 1989: 36).

Limitations of and lessons from the garbage can model. The garbage can model has been used effectively to describe the decision-making and policy-formation processes. Having a clearer picture of the complexity is of enormous benefit to people involved in the policy process. Nonetheless, the model can be faulted for a decidedly structural rather than agent-oriented approach to these processes and a resulting sense that these structural components will play out and that participants can do little except be ready and try to be in the right place at the right time. This way of describing decision making is particularly limiting to unorganized interests (Weiner 1976), such as members of the public whose jobs are not related to their ongoing participation in the policy process and who may find it difficult to remain involved once visible choice points have passed (Sabatier 1975). It does, however provide insights—into the nonlinear and continuous nature of policy formation; the potential for entrepreneurs (including public managers, elected officials, and technical experts) to make connections between streams; the importance of creating and recognizing choice opportunities; and the ambiguous nature of information and the need to generate opportunities for interpretation of policy problems and potential policy solutions.

Policy Formation as a Continuous Process

Given the complexities of making decisions, illustrated by the garbage can model, what are the options for policy makers looking for ways to address these difficult circumstances and for managers trying to move forward? How can decision makers address multiple issues, such as environmental protection, economic development, and poverty alleviation, in ways that enable people to take actions that make a difference?

The emphasis needs to move away from decision making toward action as part of a continuous policy process. The decision-making frame focuses attention on the pronouncement of a policy; at best it produces an understanding of policy formation as a staged process in which decisions are revisited and adjusted. The concern with policy formation, however, cannot stop with the pronouncement of a policy. It must also be possible to implement the policy (Brunsson 1985; Moore 1995; Feldman and Khademian 2007). Extending the understanding of policy formation in this critical direction requires thinking about how the policy-formation process influences implementation and, indeed, viewing policy formation and implementation as continuous rather than separate stages.

This section presents two models that have emerged as ways of incorporating concerns with implementation into the policy-formation process in the complex context in which policy is inevitably formed. One, "adaptive management," is oriented to enabling action in the face of uncertainty; the other, "inclusive management," is oriented to enabling action in the face of ambiguity. Both uncertainty and ambiguity are almost always present in decision and policy-formation processes (Moe 1990). These concepts have been discussed in terms of beliefs about cause and effect relations and preferences about possible outcomes (Thompson and Tuden 1959; Thompson 1967). The models are combined here in order to propose a way of promoting policy that facilitates dealing with both dimensions.

Uncertainty often relates to the dimension of beliefs about cause and effect (there may also be uncertainty about other aspects of policies and decisions). Uncertainty occurs whenever the information that would be required to resolve a question can be specified (Feldman 1989). There is often uncertainty about future effects. One cannot know what impact a new dam will have on an endangered species, but one can know what information needs to be gathered to assess this impact. Being able to specify the information needed to resolve uncertainty does not imply that the cost of obtaining the information is reasonable or even that the information is obtainable; often estimates or predictions are used.

Ambiguity—that is, situations that have many meanings (Feldman 1989)—is often related to the dimension of preferences. The multiplicity of meanings is important, because resolving multiple meanings is a fundamentally different process from resolving a lack of information. Thus, how critical it is to preserve a particular species of fish that may be endangered by a new dam and to what degree the fish should be protected are issues of ambiguity. They are matters of interpretation; the answers to these questions will vary based on one's perspective. Specific pieces of information will not resolve ambiguity. Indeed, though gathering information is often necessary in the face of ambiguity, more information often increases rather than decreases ambiguity. Processes of interpretation, which are fundamentally social processes, are necessary for resolving ambiguity (Feldman 1989). Taking action in the context of ambiguity often requires social processes that produce some consensus on a way of understanding.

Both uncertainty and ambiguity are obstacles to taking action. Under uncertainty decision makers may know what they want to accomplish, but they may not know what effects their actions will have. Under ambiguity, they may not know what effects they want their actions to have or even what effects it would be appropriate for them to have.

Both uncertainty and ambiguity are ubiquitous. If some uncertainty is resolved, other uncertainties emerge. As new uncertainties emerge, they create new ambiguities.

Uncertainty and ambiguity are also intertwined. Ambiguities are often revealed by efforts to resolve uncertainty, and how decision makers resolve uncertainties often influences how they address ambiguities. Reconceptualizing policy formation as a continuous process is necessary for understanding how policies emerge in these very real contexts.

These concepts can be illustrated with research on water sector reform in Ceará, Northeast Brazil (Ballestero 2004, 2006). In the mid-1980s, the State of Ceará, began to change the way water was managed in one of the most populated semi-arid regions of the world. The new, centrally defined water policies and changes in regulatory design and jurisdiction were premised on participation, decentralization, and integration of diverse user needs at the local level, where decisions about the volume of water to be discharged from any given reservoir would be made and permits would be issued for human consumption, industrial use, and farming use of the water. The policy also introduced economic valuation of water as a principle shaping everyday decisions about appropriate water use; participation of water users; and incentives for some productive activities and not others.

The case illustrates the continuous nature of policy formation and the fluid nature of participation. While the formal changes by the state clearly provided a catalyst for the changes that eventually took place in the practice of regulating water use in Ceará, ongoing, localized actions were critical in determining the actual adaptations in water use and the potential use of these opportunities to promote more-inclusive practices.

Managing Uncertainty

Adaptive management is a model that is oriented primarily to enabling action in the face of uncertainty. Participants may have a clear sense of what they would like to accomplish, but the impact of actions taken to accomplish the goal are not clear. Adaptive management has been used in a variety of fields; it is most common in the field of environmental policy (Holling 1978). This model of policy formation promotes the use of quasi-experiments that take place as an ongoing part of the policy process (Jacobs and Wescoat 2002). It involves taking action while there is still considerable uncertainty about the consequences, but designing actions so that they can be monitored and adjusted as their effects become more clearly understood. As Lee (1999) notes, "Management policies are designed to be flexible and are subject to adjustment in a social learning process" (cited in National Research Council 2004: 20).

While there is no formula for adaptive management, elements generally include the following (National Research Council 2004):

- Management objectives that are regularly revisited and accordingly revised
- A model (or models) of the system being managed
- A range of management choices
- The monitoring and evaluation of outcomes

- A mechanism for incorporating learning into future decisions
- A collaborative structure for stakeholder participation and learning.

The basic idea of this model is that an action is tried and its outcome evaluated in order to determine what to try next. Policy is formed through a process of experimentation, the adoption of successful experiments, and continued experimentation. An assumption of this model is that the desirability of an outcome will be relatively clear once the outcome occurs. Whether one likes what has occurred is an issue of ambiguity; for this one needs a model that speaks to the bringing together of multiple perspectives—that is, a collaborative structure for stakeholder participation and learning. The inclusive model, developed next, addresses this issue more directly by characterizing specific projects as opportunities for the ongoing creation of communities of participation.

Managing Ambiguity

The inclusive model recognizes that managing ambiguity is fundamentally a social process. It is based on understanding the intrinsic ambiguity revealed by many policy-making efforts and the importance of combining multiple perspectives in problem-solving efforts in a community of participation (the people and organizations involved in a process of policy making and implementation and the way they interact) in order to deal constructively with that ambiguity. In each policy-making or implementation process, a community of participation is created. The creation of this community may be purposeful or simply consequential. A mandate, such as the centrally defined water policies in Ceará, can specify a breadth of participation in a policy process; managers and other participants may take actions to broaden the base of participation to include differing perspectives and experiences or to narrow the range of participation and exclude varied perspectives and experiences. But communities of participation are created and recreated even without the intentional efforts of legislators, managers, or other participants; people and organizations flow in and out of policy processes as choice opportunities arise. Communities of participation become increasingly inclusive as the categories of participants who view one another as having a legitimate role to play in a joint problem solving process increase.

The creation of an inclusive community of participation (in which a wide range of perspectives involved in a process of policy making and implementation is viewed as having a legitimate role to play) is an essential outcome of inclusive management. It is achieved through the engagement of participants in specific projects (Feldman and Khademian 2000, 2007; Khademian 2002). Where traditional management models may see building community capacity as a by-product of solving policy problems, this model places capacity building in the foreground. Building capacity becomes the primary goal and projects a way of pursuing that goal. Capacity building is not something that is done instead of projects, it is something that is done through projects.

A Model of Continuous Policy Formation

Combining the models of adaptive and inclusive management creates a model of continuous policy formation. Every action taken (for example, launching experiments; assessing those experiments; deciding to continue, to stop, or to modify them) not only creates some intended policy intervention, it also contributes to the creation and recreation of a community of participants in the policy process. The actions taken will facilitate the breadth and form of participation for existing and potential members of the community, the opportunities to have perspectives legitimated and to be influenced by others, and the possibilities for connecting with and learning from others. The community of participation, in turn, has implications for the choice opportunities that arise, the information that is brought to bear on any choice opportunity, and the interpretation and legitimacy of assessments of any policy effort. It is the community of participation that determines who will take what actions and in what ways, who will assess those actions, and by what standards.

The relation between adaptive management and the community of participation of the inclusive model can be depicted graphically (figure 3.2). The L-shaped arrows connect the actions undertaken as part of adaptive management to respond to the uncertainty in a policy situation. The double-headed arrows depict the relation between the actions taken and the community of participation. The character of the community of participation influences the actions taken, and the actions taken influence the nature of the community of participation. The community changes over time and in response to the policy process. Ambiguity

FIGURE 3.2
Continuous Policy Process

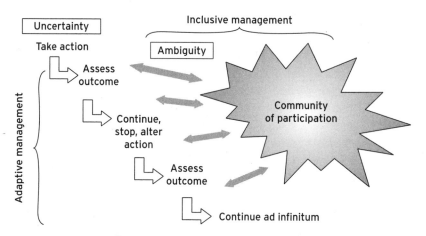

Source: Authors.

is exposed in the efforts to find actions that the community can support and engage in.

Policy formation surrounding the allocation of water from a reservoir in the small community of "Green Forest" in Ceará, Northeastern Brazil, illustrates the interplay between uncertainty and ambiguity (Ballestero 2006). The process involves uncertainty, in that the right amount of water to be discharged from a reservoir in any given year and the effects of the discharge are not known. This uncertainty can be resolved by a number of means, including technical modeling of varied amounts of water discharge. Regardless of how the uncertainty is dealt with, the process also involves ambiguity, because there are different ways of valuing the water and assessing the outcomes of actions. What uses of water are most important? Are technical or experiential means of assessing potential actions more appropriate? How much risk is bearable? Such questions will be answered differently by people with different perspectives.

Passage of water reform legislation in Ceará in 1994 gave water users increased decision-making authority in managing reservoirs. In Green Forest users of the reservoir include residents, whose concerns for water allocation focus on protecting the purity and quantity of their water. They also include small to medium-size farmers who depend on irrigation, whose primary concern is ensuring an appropriate level of discharge for their farms. (As farmers are also residents, there is substantial overlap between the concerns of these two groups.) By mandate the community of participation surrounding the Green Forest reservoir also includes staff from the water agency, who provide technical estimates on the amount of water in the reservoir. By custom, the staff of the water agency determine how much water is released. The individuals who represent residents, farmers, and staff change over time.

The following vignette—a small part of this complex case—illustrates how efforts to resolve uncertainty about the amount of water to be released and the means of measuring that water expose ambiguities related to different ways of valuing water and different understandings about how much risk to bear in the use of water from a local reservoir. It shows how efforts to respond to uncertainty influence the creation of a community of participation and the ability of that community to work out these ambiguities. It also shows how the working out of these ambiguities influences the composition and nature of the community of participation.

In 2001, the water agency staff conducted a study to assess the amount of water in the aging reservoir. Because of the large quantity of eroded soil, it was difficult to determine how much water was available for use. Efforts to resolve this uncertainty revealed differences in interpretation. Technical difficulties with the measuring equipment produced estimates that the community received with great skepticism but that the water agency was willing to use as a basis for releasing water. The agency used this information to calculate an estimate for water release and decided to discharge water accordingly. Aware of residents' opposition to releasing water, the local reservoir manager asked for police backup when he started the discharge. Residents assembled to prevent the release; after the police left, they broke the valve to ensure that the water would not be released. These actions exposed ambiguity, in the form of substantial differences in the ways that water agency staff and residents valued water and what standards and actions they thought appropriate.

The community of participation at this point consisted of water agency staff and residents. The nature of this community was divisive. Water agency staff and residents were not working together to resolve their differences. Possibilities for interaction between residents and the water agency were limited by the resident perceptions of illegitimacy associated with the technical, estimated information, and by the water agency members' determination that there was enough water to release in contrast to the residents' skepticism about the quality of the information and their reluctance to bear the risk of having insufficient water for residential uses.

Over the next two years, tension between these groups persisted. A meeting was held every year in which the water agency proposed releasing water. Residents challenged the proposals, at meetings that were characterized by overt conflict. Though agreement was eventually achieved, the community of participation remained essentially divided.

The 2001 study, the attempted release of water, the reaction of residents, and the residue of accumulated tension from the 2003 and 2004 meetings formed an important backdrop to the 2005 meeting. There were also new participants, from both the community and the water agency, including the person from the water agency in charge of mobilizing residents. Residents also had a new representative, in the form of the new school principal. Both these people had the ability to alter social interactions by facilitating communication between residents and water agency staff.

The discussion began with a presentation of technical data by a manager from the water agency. At the conclusion of his presentation, he asked for questions and suggested that residents make a decision on the options he had presented. His request was met with silence. After several moments of silence, the school principal spoke up and said that he thought the community members did not want to talk about the options. His comment released a flood of pent-up complaints by community members. The water agency staff listened. When things calmed down, they described the problem from their perspective. They explained the technical problems with the study and noted that the study had been conducted by people outside the agency before the agency had been given responsibility for the reservoir.

By the end of the meeting two changes had taken place. First, residents and water agency staff agreed not to release water from the reservoir for the coming year. Second, the community of participation had changed. The complaining, listening, and explaining produced a change in the way people treated one another. Water agency staff had a better understanding of community members' standpoint, and community members had a better understanding of the position water agency staff were in. Before everyone left the room, a teacher thanked the managers for coming to their community and stressed that the community knew that the managers were not to blame for the organization's actions and that residents liked them as individuals but could not accept any chance of losing their water (Ballestero 2006).

The new community of participation involved recognizing the constraints and concerns of both technical staff and residents. This recognition resulted in a more-amicable set of relations, which formed the basis of the 2005 agreement surrounding water discharge from the reservoir. In the new community of participation, technical knowledge did not dominate political and social concerns held by local residents. As a result, both water agency staff and residents became participants in a different kind of community of participation, in which the process of resolving uncertainty about how much water to release was understood to have implications for the nature of the community.

This example illustrates the continuous processes that influence the community of participation and the actions that are taken. Actions taken expose ambiguities that require the community of participation to address them (figure 3.3). The process of addressing ambiguities creates and recreates the community of participation.

FIGURE 3.3
Continuous Policy Process: Green Forest

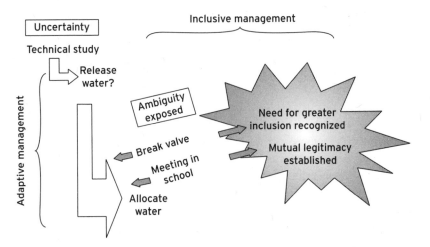

Source: Authors.

The process is a continuous one. New ambiguities emerge as marginal participants become more central and people who have not participated become participants. The way these ambiguities are dealt with will have implications for the ongoing creation of the community of practice. In Green Forest farmers, whose interest overlap to some extent but not entirely with (nonfarmer) residents, still feel that their perspective is not included and their role in the community of participation is not as relevant as those of either residents or water agency staff members. As farmers are included, more ambiguities will almost certainly emerge. Moreover, as the community of participation develops ways of resolving uncertainties about how to measure available water and how much water to release, other uncertainties will emerge. If water is released in the future, for instance, there will be questions about what this water can be used for (Ballestero 2006). These new uncertainties will reveal new ambiguities and new potential participants in the community of participation.

Conclusion

This chapter examines two ways of conceptualizing the policy-making process. The first is as a decision-making process. The models examined can be arranged on a continuum of increasing uncertainty and ambiguity. Linear models of policy formation, such as SEA, are adapted from decision-making models. They accommodate less uncertainty and ambiguity than nonlinear models.

Nonlinear models of decision making, such as the garbage can model, reveal the importance of ambiguity and the need to generate opportunities for interpretation in the policy-formation process. These revelations suggest that in order to understand how policy making works in practice, one needs to conceptualize policy making as a continuous process. Combining adaptive management with the creation of an inclusive community of participation and the consequent acknowledgment and exposure of ambiguity creates a model of policy formation as a continuous process. In this model the character of the community of participation influences the actions taken by participants; those actions in turn influence the nature of the community of participation. Efforts to resolve uncertainties and take action can reveal ambiguities; the ability to resolve ambiguities is influenced by the constitution and the nature of the community of participation that is formed in part through efforts to resolve uncertainties.

The case study illustrates the important role that managers of government agencies can play in shaping the community of participation. A rich literature on participation and community empowerment documents efforts to solve problems without input from government agencies (Rosenau and others 1992; Peters and Pierre 1998; Fung 2003; Boyte 2005). Although this type of action is often necessary, it is important to address ways in which government agencies can help. Public managers often have responsibility for projects that can be used to promote an inclusive community of participation. These projects can be opportunities to set up and facilitate processes that promote information exchange and the ability to understand the perspectives of other participants. They can also be opportunities to create relationships that enable people to work together to address problems (Feldman and Khademian 2002). Efforts to accomplish governance without government fail to take advantage of these opportunities.

As development organizations continue to find ways to bring environmental considerations to the formation of policies, it is useful to view tools such as SEA as part of a continuous process rather than as a means to intervene "upstream" in a linear process. The actions associated with SEA can reveal ambiguity within a community of participation. They can focus attention on the social processes involved in addressing ambiguity and the importance of building communities of participation that can constructively address ambiguity in the pursuit of effective action.

Note

1   The term *policy formation* is used rather than *policy formulation*. An understanding of policy making as stages in a linear process evokes the concept of formulation, or a formula for making policy. In contrast, the term *formation* does not imply a particular structure or steps toward policy but instead allows for the complexity, simultaneity, and often spontaneity of choice opportunities and implementation efforts.

References

Allison, Graham. 1971. *The Essence of Decision: Explaining the Cuban Missile Crisis.* Boston: Little, Brown.

Allison, Graham, and Philip Zelikow. 1999. *Essence of Decision,* 2nd ed. New York: Longman.

Ballestero, Andrea. 2004. "Institutional Adaptation and Water Reform in Ceará: Revisiting Structures for Social Participation at the Local Level." Master's thesis, University of Michigan, Ann Arbor.

———. 2006. "Water Policies, Resistance and the Making of Social Change: Scenes from Northeast Brazil." Paper presented at the 26th International Congress of the Latin American Studies Association, San Juan, Puerto Rico.

Boyte, Harry. 2005. "Reframing Democracy: Governance, Civic Agency, and Politics." *Public Administration Review* 65 (5): 536–46.

Bozeman, Barry. 1986. "The Credibility of Policy Analysis: Between Method and Use." *Policy Studies Journal* 14 (4): 519–39.

Brunsson, Nils. 1985. *The Irrational Organization: Irrationality as a Basis for Organizational Action and Change.* New York: John Wiley and Sons.

Cohen, M. D., J. G. March, and J. P. Olsen. 1972. "A Garbage Can Model of Organizational Choice." *Administrative Science Quarterly* 17 (1): 1–25.

Cyert, R. M., and J. G. March. 1963. *A Behavioral Theory of the Firm.* Englewood Cliffs, NJ: Prentice-Hall.

Dalal-Clayton, Barry, and Barry Sadler. 2005. *Strategic Environmental Assessment: A Sourcebook and Reference Guide to International Experience.* London: International Institute for Environment and Development.

Feldman, Martha. S. 1989. *Order without Design*. Stanford, CA: Stanford University Press.

Feldman, Martha S., and Anne M. Khademian. 2000. "Management for Inclusion: Balancing Control with Participation." *International Public Management Journal* 3 (2): 149–68.

———. 2002. "To Manage Is to Govern." *Public Administration Review* 62 (5): 541–54.

———. 2007. "The Role of the Public Manager in Inclusion: Creating Communities of Participation." *Governance* 20 (2): 305–24.

Feldman, Martha S., and James G. March. 1981. "Information as Signal and Symbol in Organizations." *Administrative Science Quarterly* 26 (2): 171–86.

Flyvbjerg, Bent. 1998. *Rationality and Power*. Chicago: The University of Chicago Press.

Fung, Archon. 2003. "Associations and Democracy: Between Theories, Hopes and Realities." *Annual Review of Sociology* 29: 515–39.

Gersick, C. J. 1991. "Revolutionary Change Theories: A Multilevel Exploration of Punctuated Change Paradigm." *Academy of Management Review* 16 (1): 10–36.

Halperin, Morton H. 1974. *Bureaucratic Politics and Foreign Policy*. Washington, DC: Brookings Institution.

Hochschild, Jennifer L. 1984. *The New American Dilemma: Liberal Democracy and School Desegregation*. New Haven, CT: Yale University Press.

Holling, C. S., ed.1978. *Adaptive Environmental Assessment and Management*. New York: John Wiley and Sons.

Hug, Simon. 1999. "Nonunitary Actors in Spatial Models: How Far Is Far in Foreign Policy?" *Journal of Conflict Resolution* 43 (4): 479–500.

Isenberg, Daniel. 1984. "How Senior Managers Think." *Harvard Business Review*, November/December: 92–98

Jacobs, Jeffrey W., and James L. Wescoat, Jr. 2002. "Managing River Resources: Lessons from Glen Canyon Dam." *Environment* 44 (2): 8–19.

Khademian, Anne M. 2002. *Working with Culture: How the Job Gets Done in Public Programs*. Washington, DC: CQ Press.

Kingdon, John. 1995. *Agendas: Alternatives and Public Policies*, 2nd ed. New York: Harper Collins.

Lee, K. N. 1999. "Appraising Adaptive Management." *Conservation Ecology* 3 (2): 3.

Lindblom, Charles. E. 1959. "The Science of Muddling Through." *Public Administration Review* 19: 79–98

Lindblom, Charles E., and David Cohen. 1979. *Usable Knowledge*. New Haven, CT: Yale University Press.

Lynn, Lawrence E., Jr. 1978. "The Question of Relevance." In *Knowledge and Policy: The Uncertain Connection*, ed. L. E. Lynn, 12–22. Washington, DC: National Academy of Sciences Press.

March, James G. 1978. "Bounded Rationality, Ambiguity and the Engineering of Choice. *Bell Journal of Economics* 9 (2): 587–608.

———. 1991. "Exploration and Exploitation in Organizational Learning." *Organization Science* 2 (1): 71–87.

March, James G., and Johan P. Olsen. 1976. *Ambiguity and Choice in Organizations*, Bergen Norway: Universitetsforlaget.

———. 1989. *Rediscovering Institutions: The Organizational Basis of Politics*. New York: Free Press.

March, J. G., and H. A. Simon. 1958. *Organizations*. New York: Wiley.

Mintzberg, Henry. 1994. "The Fall and Rise of Strategic Planning." *Harvard Business Review* (January–February): 107–14.

Moe, Terry. 1990. "The Politics of Structural Choice: Toward a Theory of Public Bureaucracy." In *Organizational Theory from Chester Barnard to Present and Beyond*, ed. Oliver E. Williamson. New York: Oxford University Press

Moore, Mark. 1995. *Creating Public Value: Strategic Management in Government.* Cambridge, MA: Harvard University Press.

National Research Council. 2004. *Adaptive Management for Water Resources: Project Planning.* Washington, DC: National Academies Press.

Peters, Guy P., and John Pierre. 1998. "Governance without Government? Rethinking Public Administration." *Journal of Public Administration Research and Theory* 8 (2): 223–43.

Rosenau, James N., Ernst-Otto Czempiel, Steve Smith, Thomas Biersteker, Chris Brown, Phil Cerny, Joseph Grieco, A. J. R. Groom, Richard Higgott, G. John Ikenberry, Caroline Kennedy-Pipe, and Steve Lamy, eds. 1992. *Governance without Government.* Cambridge: Cambridge University Press.

Sabatier, Paul. 1975. "Social Movements and Regulatory Agencies: Toward a More Adequate and Less Pessimistic Theory of Clientele Capture." *Policy Sciences* 6 (3): 301–42.

Seidman, Harold. 1980. *Politics, Position, and Power.* New York: Oxford University Press.

Simon, Herbert. 1957a. *Administrative Behavior: A Study of Decision Making Processes in Administrative Organization*, 2nd. ed. New York: Macmillan.

———. 1957b. *Models of Man, Social and Rational.* New York: John Wiley and Sons.

Steinbruner, John. 1974. *The Cybernetic Theory of Decision.* Princeton, NJ: Princeton University Press.

Stevenson, William, Jone Pearce, and Lyman Porter. 1985. "The Concept of 'Coalition' in Organization Theory and Research." *Academy of Management Review* 10 (2): 256–68.

Stockholm Environment Institute. 2003. "Research and Advice on Strategic Environmental Assessment." November. http://www.sei.se/policy/SEA.pdf.

Thompson, James D. 1967. *Organizations in Action.* New York: McGraw-Hill.

Thompson, James D., and Arthur Tuden. 1959. "Strategies, Structures and Processes of Organization Decision." In *Comparative Studies in Administration*, eds. James D. Thompson, Peter B. Hammond, Robert W. Hawkes, Buford H. Junker, and Arthur Tuden. Pittsburgh, PA: University of Pittsburgh Press.

Vaughan, Diane. 1997. *The Challenger Launch Decision.* Chicago: University of Chicago Press.

Walters, Carl, and C. S. Holling. 1990. "Large-Scale Management Experiments and Learning by Doing." *Ecology* 71 (6): 2060–68.

Weick, K. E., and R. E. Quinn. 1999. "Organizational Development and Change." *Annual Review of Psychology* 50: 361–86.

Weiner, Stephen S. 1976. *Ambiguity and Choice in Organizations*, eds. James G. March and Johan P. Olsen. Bergen, Norway: Universitetsforlaget.

Weiss, Carol H. 1977. "Research for Policy's Sake: The Enlightenment Function of Social Science Research." *Policy Analysis* 3 (4): 531–45.

———. 1978. "Improving the Linkage between Social Research and Public Policy." In *Knowledge and Policy: The Uncertain Connection*, ed. L. E. Lynn. Washington, DC: National Academy of Sciences Press.

————. 1980. "Knowledge Creep and Decision Accretion." *Knowledge: Creation, Diffusion, Utilization* 1 (3): 381–404.

Wildavsky, Aaron. 1979. *The New Politics of the Budgetary Process*, 3rd ed. Boston: Little, Brown.

Wilensky, Harold L. 1967. *Organizational Intelligence: Knowledge and Policy in Government and Industry*. New York: Basic Books.

Wilsford, David. 1994. "Path Dependency, or Why History Makes It Difficult, But Not Impossible to Reform Health Care Services in a Big Way." *Journal of Public Policy* 14 (3): 251–83.

Wolfers, Arnold. 1962. *Discord and Collaboration: Essays on International Politics*. Baltimore, MD: Johns Hopkins University Press.

World Bank. 2002. "Strategic Environmental Assessment (SEA): A Structured Learning Program at the World Bank." SEA SLP Progress Note, June 27, Washington, DC.

CHAPTER 4

Toward Environmental Priority Setting in Development

Richard D. Morgenstern

THE INTEGRATION OF ENVIRONMENTAL ISSUES into the mainstream policy arena involves issues of sustainability, government decentralization, efficiency, effectiveness, and transparency. The push for sustainability reflects a concern that environmental considerations should be a part of all government policies, not just those formally designated "environmental"; gone are the days when a single minister or cabinet official at the national level was responsible for all environmental programs. Mitigating widespread environmental problems requires a broad array of actions, touching on many sectors of the economy. The daunting and politically difficult task of setting priorities across this array of actions is the first element in devising cost-effective strategies for addressing environmental problems.

Because of the difficulty of the task, a formal government process in which priorities are made explicit is not common. In practice, environmental priorities can be influenced by public clamor; cultural, historical, institutional, and political factors; development agency priorities; international agreements; judicial decisions; and the results of technical studies.

This chapter examines alternative approaches to environmental priority setting that emphasize formal analytical/quantitative techniques. The focus is on the techniques used to evaluate risks and economic damage.[1]

Richard D. Morgenstern is a senior fellow at Resources for the Future, in Washington, DC.

The chapter is organized as follows. The next section presents background information on the use of these techniques, along with selected critiques of both the risk and economic approaches. The second section reviews the analytical framework and on-the-ground experiences with comparative risk assessment (CRA) in both the developed and developing worlds. The third section focuses on the methods used to conduct economic damage assessments and reviews the recent experiences with this approach in 10 developing countries.[2] The last section examines the similarities and differences between the two methodologies and draws conclusions that can help refine future efforts to mainstream environmental issues.

Background

Environmental protection policies enjoy strong and growing support worldwide. Public opinion surveys; efforts by legislators and government officials to enact and implement environmental laws; and the array of programs and financial support from development agencies all demonstrate the strong international commitment to reducing risks to human health and the environment. Increasingly, environmental objectives are being integrated into other policy areas such as energy, agriculture, and tourism. Over the past decade, the World Bank and other institutions have put considerable effort into integrating both the risk and economic approaches into their country and strategic environmental assessments in developing countries. Increasingly, economic damage assessments are addressing a range of "green" issues, and attention is turning to nonurban and nonindustrial sectors as areas of interest.

Concurrent with the growing demand for greater environmental protection, serious concerns have arisen in many countries about whether the public agencies in charge of environmental management are directing scarce resources toward the most serious problems. The notion that resources devoted to environmental protection should be allocated to the most serious problems has broad appeal. When augmented with information on mitigation costs, the findings on risk and economic damage can be used to make sound environmental management decisions.

The methods used to prioritize environmental problems are sometimes controversial. Environmental advocates, for example, often see the use of such techniques as a distraction; they prefer to focus on increasing the size of the pie rather than fighting over the slices. Some government officials see risk- and economic-based criteria as undermining the normal workings of the political process or as inconsistent with legislative intentions. They see a "tyranny of experts," whose controversial scientific and pseudoscientific methods are used to create a false sense of objectivity and to shift emphasis away from traditional practices and beliefs, often ignoring difficult ethical and social choices. Other observers see the use of these techniques as overemphasizing the easily measured indicators of human health or environmental quality at the expense of forests, coastal management, and other hard-to-measure "green" issues.

Even in the more data-rich area of health analysis, many challenges remain. How does one compare the risk of a small drop in intelligence for children exposed to lead versus the risk of heart attacks among the elderly induced by elevated levels of fine particulates in the ambient air? Which is worse, one person dying or 10,000 people being slightly impaired? Does it matter if the person dying is a child or a smoker? How should ecological risks be compared with risks to human health? Critics contend that scientists are expert only at determining probabilities, that the public or its representatives should be asked to contribute their expertise to the process of valuation.

Notwithstanding some resistance to risk- and economic-based techniques, traditional approaches are sometimes challenged by the discovery of new information. Such information often comes from a distant country or region with new scientific evidence on illness or death associated with particular types of pollution. Sometimes the information arises out of local experiences; occasionally it comes from tragedies, such as the disasters in Seveso, Italy, or Bhopal, India, in the 1980s. Whatever its source, new information can trigger a rethinking of traditional policies and practices for environmental protection.

The growing interest in analytical techniques for priority setting signals a major shift from the initial stages of environmental policies, which date back to the 1970s in many countries. Those earlier stages emphasized the existence of problems rather than their magnitude and were often rooted in moral indignation directed at the behavior of polluters.

Of course, this early environmentalism can claim many successes. The U.S. Environmental Protection Agency (EPA) estimates that as a result of the Clean Air Act (originally enacted in 1970), the number of premature deaths among people 30 and older fell by almost 200,000 between 1970 and 1990, and the number of cases of congestive heart failure, chronic bronchitis, and asthma attacks fells by hundreds of thousands; ecological impacts were also reduced and agricultural yields improved (U.S. EPA 1997). Yet complaints of high costs and inefficiency have accompanied even these well-documented accomplishments. For other environmental issues—such as water pollution, where the results are less readily quantified—complaints have been louder. This, in turn, has spawned efforts to introduce more scientific and economic rigor into the policy process.

Despite the growing support for analytic/quantitative techniques in many public policy areas, such as health care and education, hostility to their use in environmental decision making remains an issue in many circles. Critics argue that the disciplines of risk and economic analysis are too narrow to deal with the moral and ethical issues associated with environmental policy and that the paradigm of individual preferences does not provide a broad enough basis on which to make environmental decisions. Concerns have also been raised about the possibility of manipulation of information. It is sometimes claimed that because of the primitive level of knowledge about the physical effects of pollution—not to mention some of the issues associated with the valuation of benefits and costs—the risk and economic analysis of environmental regulation is too readily manipulated. Yet, as Herman Leonard and Richard Zeckhauser (1986: 3) argued more than two decades ago:

> Any technique employed in the political process may be distorted to suit parochial ends and particular interest groups. [Economic] analysis can be an advocacy weapon, and it can be used by knaves. But it does not create knaves, and to some extent it may even police their behavior. The critical question is whether it is more or less subject to manipulation than alternative decision processes. Our claim is that its ultimate grounding in analytic disciplines affords some protection.

Notwithstanding some legitimate concerns about the techniques, there is growing support for the use of analytical studies to help improve the allocation of society's resources and engender an understanding of who benefits and who pays for any given policy. Such studies can encourage transparency and accountability in the decision-making process and provide a framework for consistent data collection and identification of gaps in knowledge. Economic analyses have the additional advantage of allowing for the aggregation of many dissimilar types of damages (such as those on health, visibility, and crops) into one measure of net damages/benefits expressed in a single metric.

The fields of risk analysis and environmental economics have expanded rapidly over the past several decades. A distinguished group of economists, including Nobel laureate Kenneth Arrow, has identified three major functions of economic analysis in the regulation of health, safety, and the environment: arraying information about the benefits and costs of proposed regulations; revealing potentially cost-effective alternatives; and showing how benefits and costs are distributed (geographically, temporally, and across income and racial groups, for example) (Arrow and others 1996).

A number of studies document how risk and economic analyses have been used to support environmental management efforts. In the United States, for example, such analyses have helped produce and accelerate stronger regulation of lead in gasoline; promoted more-stringent regulation of lead in drinking water; led to stronger controls on air pollution at the Grand Canyon and the Navaho Generating Station; and supported the use of cleaner motor vehicle fuels (Morgenstern 1997). In Europe, where there are fewer requirements to conduct such studies, their use is also growing.[3]

Risk and economic analyses have also been used to evaluate the equity and redistributive aspects of environmental policies. There are numerous examples of such evaluations in the developed world; the techniques have been used to examine equity issues in the developing world as well. In Bogotá, Colombia, for example, analytical studies have documented that the parts of the city with the dirtiest air are also the poorest (Blackman and others 2005). These studies produce an additional rationale for improving air quality: enhancing the well-being of low-income groups.

In sum, while there is no perfect method for establishing environmental priorities, there are many reasons to believe that risk- and economic-based techniques have an important role to play. In the United States, much debate has occurred on how to expand the use of use these techniques in real-world decision settings. While some would seek to create a more-formal structure, the predominant view is that such techniques are best used as tools rather than rules for decision making. In the developing world, these techniques are increasingly used in Strategic Environmental Assessments (SEAs) undertaken by national governments, development agencies, and others to reform and strengthen environmental management systems.

Comparative Risk Analysis

Comparative risk analysis (CRA) is a systematic procedure for evaluating the environmental problems affecting an area. It typically incorporates elements of both risk assessment and risk management. In the United States this approach is used by the federal government and by subnational entities in more than 50 localities. Internationally, more than a dozen countries have developed major projects based on this methodology.[4]

CRA provides a systematic framework for evaluating environmental problems that pose different types and degrees of risk to human health and the environment.[5] The basic premises of CRA are that risk provides an objective measure of the relative severity of different environmental problems and that risk reduction provides a metric for organizing and evaluating efforts to address the problems.

CRA has two stages, risk assessment and risk management. In the risk-assessment stage, the environmental problems facing an area are identified, evaluated, and compared, with the aim of ranking the problems based on the risks they pose. In some cases a single risk metric is employed. In other cases different rankings are developed for different risk categories (cancer versus noncancer health risks, health versus ecological risks. The ranking process involves assembling and analyzing relevant data on environmental problems (including information from scientific risk analyses) and using structured judgments to fill gaps in the data. Although the risk-ranking process is scientifically oriented, it relies heavily on value judgments. The hazards to be considered in the risk assessment, how "risk" is to be measured, how different risks should be weighted, and how uncertainty should be treated are matters that often involve local values and social choice.

Although not incorporated into all CRAs, risk management involves the development and assessment of initiatives to address the key risks identified in the risk-assessment phase. The considerations in this stage extend beyond risk assessment to include economic, technical, institutional, legal, and political factors. CRA was originally developed by the U.S. EPA and published as an agency report entitled *Unfinished Business* (U.S. EPA 1987). Although *Unfinished Business* did not immediately lead to a major reallocation of resources at EPA or radically change public perceptions of environmental risks, it did help broaden thinking in the policy community on the need to prioritize efforts on the basis of the potential for risk reduction.

Since publication of *Unfinished Business*, more than half of all U.S. states and more than 50 localities have employed CRA to identify and address important environmental issues. Internationally, CRAs have been conducted in cities in more than a dozen developing countries and transition economies, including Bangkok, Cairo, and Quito. There is evidence that risk-based studies have influenced many choices about resources that have led to better health and ecological protection (Sessions and others 1997; Keane and Cho 2000).

Feldman, Ralph, and Rulth (1996) identify four principal goals of CRAs:

- Involve the public in the priority-setting process, and identify and incorporate their concerns.
- Identify the greatest environmental threats and rank them accordingly.
- Establish environmental priorities.
- Develop action plans and strategies to reduce risks.

Individual CRAs vary in their methodological approaches and in the extent to which they fulfill these objectives. Given the diversity in both the scope and methods of CRAs, choices must be made about the elements selected as the basis of any review of these analyses. Konisky (1999) identifies three elements of particular importance:

1. *Environmental problem list.* This list typically includes a core set of pollution problems typically managed by environmental agencies (air pollution, surface water pollution, drinking water contamination, hazardous and nonhazardous waste). Increasingly, particularly in developing countries, the problem list is expanding to include issues of broader social concern, such as traffic accidents, deforestation, and occupational exposures to toxic substances. Although *Unfinished Business* included 31 problem areas, most CRAs focus on no more than a dozen.

2. *Criteria for evaluating problems.* Across all CRAs, risk provides the common denominator for comparing the disparate impacts of different environmental problems. Significant differences exist, however, in the types of risks CRA project sponsors have addressed (health and ecological risks, risks to economic well-being, and risks to the general quality of life). Even larger differences exist in

the criteria used to measure the magnitude of these endpoints (lives lost, numbers of cases of illnesses, extent of ecological damage, recovery time, monetized economic losses). Some of the criteria allow for strict quantitative estimates of harm or risk. Others are more qualitative.

3. *Ranking.* The ranking process can be thought of as a continuum ranging from a narrow process heavily determined by scientific estimates of adverse effects to a broad framework that incorporates multiple values, including public preferences. A CRA can fail by going too far in either direction. Technical analyses of adverse effects are a critical element in a CRA, but if they are the only element, the CRA is likely to have only limited impact. If the CRA incorporates all relevant elements, including political factors, and gives them the same weight they are given in the decision-making arena, it will wind up maintaining the status quo and have no value added.

No purely scientific methodologies exist for comparing very diverse types of risks. Typically, CRAs avoid cross-category comparisons, focusing instead on within-category comparisons of health, environmental, and welfare issues. While this is an attractive approach in many respects, it has a number of disadvantages:

- It provides no overall ranking and thus might not serve the purpose for which the CRA was intended.
- For some types of risks, it may be difficult to separate out the most important effects (some types of air pollution cause acute and chronic health effects, for example, as well as environmental welfare impacts).
- Even within certain categories, it may be difficult to compare different types of risks (asthma versus lung cancer, for example).
- Because environmental risks are not expressed in monetary terms, direct comparisons cannot be made with other economic indicators that could be helpful in allocating public and private resources.

It is important to focus on who is doing the ranking. Slovic (1987) and others suggest that while individuals trained in science and related fields generally employ formal risk assessment to draw conclusions about relative risk, most ordinary citizens rely instead on intuitive judgments. It is therefore not surprising that experts and the general public tend to rank the same environmental problems differently (Morgenstern and Sessions 1988; U.S. EPA 1990b).

The existence of large differences in public and expert risk perceptions highlights the normative issue of who in society should be doing the ranking. *Unfinished Business* relied exclusively on the expert judgments of government scientists and program managers. More recently, public involvement in the process has increased, in the United States and internationally. Most CRAs are now overseen by multidisciplinary working groups that include technical experts and interested individuals from institutions concerned with environmental management, including nongovernmental organizations (NGOs), businesses, universities, and the general public.

Thirteen international CRAs have been conducted since 1990, all of them in developing countries or transition economies (table 4.1).[6] Examination of these analyses reveals both similarities and differences in approaches. The number of problem areas considered in the international studies ranges from 3 to 16—far fewer than the 31 problem areas considered in *Unfinished Business*. Like *Unfinished Business*, all the studies employed a heterogeneous rather than a homogenous classification scheme (such as a pollutant, source, pathway, or receptor to define the problems). All of the studies ranked problems based on their relative health risks; a few also developed separate rankings based on ecological risks and risks to quality of life.

The groups developing the rankings differed substantially across projects. At one extreme, the risk rankings were developed by a small group of technical experts (generally less than 10 people). At the other extreme, several projects developed their risk rankings through a broad participatory process involving representatives of the general public as well as scores of experts from both government and NGOs. A few projects involved only consultants, several involved only government officials, most were multisectoral. Nearly all CRA risk-ranking processes were informed by extensive collection and analysis of technical information.

About half the CRAs proceeded beyond the risk analysis to a risk-management phase in which priorities, action plans, or initiatives were developed while considering a range of management factors as well as risk reduction. This approach differs from *Unfinished Business,* which was strictly a risk-ranking exercise.

Some of the international CRAs incorporated novel features not widely used in the United States. The USAID—sponsored CRA conducted in Quito, for example, used ethnographic methods (focus groups, structured observations, in-depth interviews). Researchers studied behaviors that affected exposure to environmental pollutants (how much drinking water they consumed, how they stored their water, whether or not they boiled the water before drinking it) in order to replace the typical default exposure assumptions with estimates reflecting local practices. They also studied people's attitudes in order to reflect local values in the risk assessment. In the local culture, for example, is an illness of more concern if it affects a child, a working adult, or an elderly person? What sorts of "quality of life" risks should be evaluated?

Despite the variety of methods used in the international CRAs, it is nonetheless useful to try to compare the risk rankings across studies (table 4.2). Is there uniformity in the relative risk rankings across different settings? Are certain environmental problems consistently found to be among the most serious?

Although the state and local CRAs conducted in the United States considered fewer problems than did *Unfinished Business,* the definitions of the environmental problems used in these projects were quite similar to the *Unfinished Business* definitions. The same cannot be said about CRAs conducted in other countries.

TABLE 4.1
Comparison of International Comparative Risk Analyses

Study Location	Number of Problem Areas	Types of Risk for Which Rankings Were Developed	Ranking or Categorizing	Project Participants	Data Collection and Analysis	Notable Features of Risk-ranking Method	Risk Management Activities Performed
Troyan, Bulgaria (ICLEI 1995)	At least 4	Health, ecological, quality of life	Categorizing	Mixed	Some	Technical committee identified list of environmental problems. Problems were divided into high, medium, and low risk based on risk analysis and public input.	Yes
Central America (USAID 1996)	7	Health, ecological, quality of life	Categorizing	Mixed. Project staff did initial risk ranking; multisectoral committee adjusted it	Extensive	Different criteria and scoring systems were used for each type of risk. Identical procedures were used in six countries and for region as a whole.	Yes
Taiwan (China) (Morgenstern, Shih, and Sessions 2000)	15	Cancer, noncancer, ecological, welfare	Some ranking, some categorizing	Government and nongovernment experts	None	Experts identified and ranked environmental problems based on their own judgment.	No
Silesia region, Czech Republic and Poland (U.S. EPA 1992, 1994)	6	Health, ecological	None	Government and nongovernment experts	Extensive	Analysis examined severity and scale of ecological as well as human health risks.	Yes

(continued)

TABLE 4.1
Comparison of International Comparative Risk Analyses (continued)

Study Location	Number of Problem Areas	Types of Risk for Which Rankings Were Developed	Ranking or Categorizing	Project Participants	Data Collection and Analysis	Notable Features of Risk-ranking Method	Risk Management Activities Performed
Quito, Ecuador (USAID 1993)	9	Health only	Categorizing	Consultants	Extensive	Analysis drew on quantitative risk assessment and health outcome data, site-specific ethnographic study, and explicit scoring of problems based on probability and severity.	No
Cairo, Egypt, Arab Rep. of (USAID 1994)	16	Health only	Categorizing	Consultants	Extensive	Estimated incidence and severity index was used to rank problems.	No
Ahmedabad City, India (USAID 1995)	11	Health, ecological, quality of life	Categorizing	Government and nongovernment experts	Extensive	Risk assessment evaluated only aggregated human health risks; focused on residual risk from current controls to guide future action plans.	Yes
West Bengal, India (Ghosh, Bose & Associates 1997)	As many as 10	Health, ecological, quality of life	Categorizing	Nongovernment experts	Extensive	Integrated risk assessment and participatory methods were based on health and ecological impacts and quality of life.	Yes

Location (source)	Number	Types of risk	Ranking/categorizing	Participants	Extent	Comments	Public involvement
Lima, Peru (Sessions and others 1997)	12	Health only	Some ranking, some categorizing	Government and nongovernment experts, the public	Extensive	Ranking reflected summary judgment of committee after reviewing information on incidence and severity of health effects and quality and biases in data.	Yes
Radom and Elk, Poland (ISC 1997)	At least 3	Health, ecological	Categorizing	Mixed	Some	Two cities identified environmental problems based on relative risk and public opinion. Rankings were based on extensive discussion among all members of study group	Yes
Zilina, Slovak Republic (U.S. EPA 1997)	Up to 6	Health, ecological	Categorizing	Mixed	Some	Work groups were established for six problems. Risk ranking is very qualitative.	Yes
Bangkok, Thailand (USAID 1990)	11	Health only	Categorizing	Consultants	Extensive	Ranking based on estimated incidence and severity index.	No
United States (U.S. EPA 1987)	31	Cancer, noncancer, ecological, welfare	Some ranking, some categorizing	Government experts	Extensive	Each work group (one for each of the four types of risk) developed its own system of criteria and scoring to rank problems.	No

Source: Author compilation.

TABLE 4.2
Health Risk Rankings in Selected Comparative Risk Analyses

Problem	UB	Taiwan (China)	Bangkok, Thailand	Quito, Ecuador	Lima, Peru	Cairo, Egypt, Arab Rep. of	Silesia, Czech Republic	Ahmedabad City, India	Central America	Troyan, Bulgaria	Zilina, Slovak Republic	Radom and Elk, Poland	West Bengal, India
Criteria air pollutants	High	High (dust) High (autotailpipe)	High (PM, lead); medium (CO); low (other criteria p.)	High	Medium	High (PM, lead); medium (ozone); medium/low (CO, SO_2)	High	High	High	High	High	High	High
Hazardous air pollutants	High	High (dioxin, gas stations)	Low			Low							
Other air pollutant	High												
Radon (indoor)	High												
Other indoor air pollutants	High	Medium		Medium	Low	Medium/low		High					
Radiation	Medium												
Ozone depletion substance	High												
Direct point source discharges	Low	High (heavy metals); medium (industrial wastewater)		Medium									
Indirect point source discharges	Medium												
Nonpoint source discharges	Medium												

Contaminated sludge	Low										
Estuaries and coastal waters	Medium										
Wetlands	Low										
Drinking water	High	High (drinking water)	Low (surface water contamination)	Medium	High (limited erratic water supply)	Medium/low (cont by chem) Medium/low (cont by microbes)	Low	High	Medium	High	High	High (water supply)
Active hazardous waste sites	Low	High (toxic substance)	Low		Low (toxic substance)	Medium/low (hazardous waste)						
Inactive hazardous waste sites	Low		Low									
Nonhazardous waste sites, municipal	Medium	Medium (solid waste)	Low	Low (solid waste)	High	Medium/low (solid waste)	Low	High		Medium/low		
Nonhazardous waste sites, Industrial	Medium											
Mining waste	Low											
Accidental toxic releases	High											
Accidental oil spills	UR											
Storage tank releases	Low											
Other groundwater contamination	NR	Medium	Low		Medium						Medium/ low	High (depletion) High (depletion)

(continued)

TABLE 4.2
Health Risk Rankings in Selected Comparative Risk Analyses *(continued)*

Problem	UB	Taiwan (China)	Bangkok, Thailand	Quito, Ecuador	Lima, Peru	Cairo, Egypt, Arab Rep. of	Silesia, Czech Republic	Ahmedabad City, India	Central America	Troyan, Bulgaria	Zilina, Slovak Republic	Radom and Elk, Poland	West Bengal, India
Pesticide residues on foods	High	High	Low	Low	Medium				High				
Application of pesticides	High												
Other pesticides risks	Medium												
New toxics chemicals	NR												
Odor	Medium												
Noise	Medium												
Endangered species	Low												
Microbiological disease		High				High							
Sea contamination					Medium								

Hazard	Risk ratings
Loss of agricultural and green land	Low
Food contamination, microorganisms	High
Traffic hazards	**Medium**
Food contamination (general)	Medium, Medium, Low
Occupational disease	High, Low
Nickel	Medium
Sewage	High, High
Surface water contamination	Low, Medium, Low
Soil contamination	Low

Source: Morgenstern, Shih, and Sessions 2000.

Definitional differences involve issues of aggregation and organizing structure. For example, *Unfinished Business* defined the problem of "criteria air pollutants" to include particulate matter, ozone, sulfur dioxide, nitrogen oxides, carbon monoxide, and lead. Other CRAs considered one or more of these air pollutants separately and ranked them individually. The Central American project defined "pesticides" as a single problem posing both human health and ecological risks. Other projects had no explicit "pesticides" problem but instead included pesticides as part of "drinking water contamination," "food contamination," or "worker exposures."

The range of the problems considered in CRAs is also important. How does one interpret the fact that in considering only six problems, the Silesia CRA omitted more than 80 percent of the problems analyzed in *Unfinished Business*? In making comparisons across CRAs, one is tempted to interpret exclusion of a problem as at least tacit acknowledgment that the omitted problem is less important than the included ones. It is also true, however, that some CRA projects have defined different definitional boundaries for "environmental problems." Should traffic hazards be considered an environmental problem? What about occupational exposures and illnesses? What about groundwater depletion (as opposed to groundwater contamination)? Not surprisingly, the organizers of a CRA are sometimes uninterested in a particular problem because it is not within the potential sphere of its responsibility.

Comparing CRAs is also hindered by substantial differences in data and methodologies. For example, all of the studies relied on similar dose-response functions in estimating health risks for key pollutants. These dose-response functions are often derived from U.S. studies. While this is not an unreasonable approach, if the true dose-response function varies substantially across countries (as a function of the age distribution or health of the population), the simplifying assumption of uniform cross-national relationships would not be appropriate.

As part of their health risk analysis, most international CRAs have adopted the practice common in U.S. analyses of aggregating and then comparing the number of deaths associated with different environmental problems. Few international CRAs distinguish between the death of an infant from diarrhea, of a working adult from an occupational accident, or of an elderly respiratory patient from air pollution. Had the CRAs used other approaches that weighted deaths among different members of society differently (using, for example, disability-adjusted life years) or through culture-specific judgments about the relative import of these different outcomes, the risk rankings of the environmental problems might have been significantly different.

Health risk, as opposed to ecological status or quality of life, was chosen as the basis for the rankings because it is the only category considered in all of the studies. The rankings in table 4.2 are disaggregated into the 27 most-frequently used of the original 31 *Unfinished Business* categories. Fourteen other rankings that were considered in the international studies but not explicitly examined in *Unfinished Business* are also included.[7]

Several observations emerge from the review of the risk rankings developed by these CRAs:

1. Air pollutants consistently rank high as health risks. Given the diverse nature of the settings studied in the CRAs—which differed in terms of economic development, ruralness, and climate—this finding is particularly important. At the same time, some differences are apparent with regard to specific air pollutants. Particulate matter and lead are found to pose high risks in virtually all settings. The ranking for sulfur dioxide is much more variable, tending to be high in areas where coal is used extensively in cities (Eastern Europe) and much lower in areas where there is little urban coal use and the predominant fuels have lower sulfur content (Bangkok, Cairo, Lima).

2. Unclean drinking water ranks as a high or medium health risk in almost all the countries studied. The reasons for this ranking differ significantly across studies. The high health risk ranking in the United States and several Eastern European communities reflects a judgment about the magnitude of the health effects caused by contaminants found in drinking water. In contrast, the CRAs of Lima and West Bengal addressed the erratic supply of clean water.

3. Problems involving toxic chemicals (hazardous air pollutants, pesticides, accidental toxic releases, radon, and radiation) tend to be ranked as greater health risks in the United States than in other countries, although dioxin and gasoline station emissions are ranked as high health risks in Taiwan (China). At the same time, several problems involving pathogens are recognized as significant problems in developing country settings. This likely reflects the shift from basic sanitation and infectious disease problems to those involving industry, vehicles, toxic substances, and other problems associated with economic development (Smith 1988). Alternatively, it may simply reflect the greater availability of information on sources of such pollutants in the United States.

4. The pollution of surface (as opposed to drinking) water is generally ranked as a medium or low health risk. In most instances exposure to contaminated surface water (by eating fish and shellfish, by swimming and engaging in other water-contact activities) was judged to be limited. While surface water used as a source of drinking water was contaminated in many locales, the treatment processes were generally thought to be acceptable. If separate rankings had been performed for ecological risks, as they were in *Unfinished Business*, pollution of surface water would likely have appeared as a greater concern.

5. Hazardous and (industrial) nonhazardous waste are ranked as medium or low health risks in all areas studied except Taiwan (China), where unmanaged toxic waste sites are believed to pose high risks. These results are consistent with the notion that few people are directly exposed to hazardous and (industrial)

nonhazardous waste. It is not entirely surprising that views may differ in Taiwan, where land area is limited and population density high.

6. Among CRAs that rank household waste, about equal numbers rank them as low and high risks. Several CRAs that rank household waste as posing low or medium health risks nevertheless mention the high risks it may pose to dump scavengers or recyclers.

7. Pesticide on food is believed to pose a high health risk in Central America, Taiwan (China), and the United States; in all other settings it ranks low on the list of health risks. Interestingly, the CRAs in Bangkok, Cairo, and Quito rank microorganisms on foods, but not pesticides, as high risks to health. These results may reflect different intensities of pesticide use across countries. Alternatively, they may indicate differences in scientific understanding of the risks associated with pesticide use.

The fact that more than 100 CRAs have been conducted over the past decade—more than a dozen of them outside the United States—reflects the widespread acceptance of the approach. At the same time, it is important to emphasize that the CRA rankings represent risk rather than priority rankings. Although a number of the studies did develop some form of priority ranking for specific control options (risk management), such rankings are not the focus of CRA comparisons. Thus, the rankings considered here do not necessarily match the rankings that should be given to environmental programs once cost, technical feasibility, and other factors are considered. These is also the nagging question of whether the observed similarities (and differences) in the results of the various CRAs reflect genuine differences in risks across the locations studied or are simply artifacts of the methods used. Corroboration by other methods would strengthen confidence in the conclusion that the observed similarities and differences in the results of the international CRAs are genuine.

Economic Damage Assessment

Economic damage assessment draws from the economics literature as well as the literature from the environmental, physical, biological, health, and epidemiological sciences. The approach is widely used in the United States and most other developed countries.

In the absence of a viable method for integrating the various health and ecological effects, the CRAs considered in the previous section stopped short of developing a single, aggregate measure of risk expressed in a single metric. Although the absence of such measure does not hinder the use of CRA results in many applications, it does hinder cross-category comparisons involving control costs. It also hinders comparisons across multiple areas of government policy and, arguably, limits opportunities to integrate environmental issues into the mainstream policy arena.

One relatively transparent means of developing an aggregate measure is assigning a monetary valuation to environmental damage. Among the many advantages of this approach is that it places environmental policies in the same framework widely used in developing, justifying, and communicating policies on most other issues.

For individual pollutants it is possible to determine the least-cost means of abatement without assigning monetary values by ranking the costs of various options for meeting environmental objectives. When multiple pollutants are involved, cost-effectiveness analysis based on physical damage is not feasible. The most commonly used way of creating a comparative framework is to convert physical damages into economic values. The monetary valuation of environmental damages involves a range of natural and social science issues. The focus here is on economic issues, because most of the natural science issues are common to both CRA and economic methods. A number of techniques have been developed to quantify and monetize the preferences and values of individuals and communities with respect to environmental quality, conservation of natural resources, and environmental health risks. Because both benefits and costs can readily be expressed in monetary terms, the analysis can provide additional support to decision makers as they allocate resources across diverse societal goals.

Costing out environmental damages involves estimating the burden of environmental pollution and natural resource damage in a number of different dimensions. Some costs are directly tied to a country's measured economic output (GDP). These include the reduced productivity of agricultural land because of erosion, salinity, or other forms of land degradation; medical treatment costs and lost work days caused by illnesses associated with environmental pollution; reduced fishery catch caused by pollution and overexploitation; and losses in tourism revenues caused by pollution and natural resource degradation. Other costs are associated with reduced well-being and quality of life. These factors, which are not explicitly included in GDP calculations, include the risk of mortality and morbidity associated with pollution and the loss of recreational quality and natural heritage caused by inadequate waste management practices, degradation of natural resources, and other factors.

When estimating the cost of environmental damage, analysts distinguish between financial and economic costs. To the extent feasible, economic costs should be applied, because they exclude transfers among groups or sectors and thus capture the net burden to society. The financial cost of health services an individual pays may be substantially lower than the true cost of providing these services, for example. The real societal cost includes the portion paid by individuals receiving the services and the portion paid by others, including the public sector. Another example is work time lost to illness or provision of care for ill family members. If the ill person or the individual providing care for an ill person does not earn income, the financial cost of time loss is zero. However, the ill person or caretaker is normally engaged in activities that are valuable to the family. The illness of a family member can impose significant burdens on the family, including reduction in the amount of time available for leisure activities. In economic analysis the opportunity cost of time (that is, the salary or fraction of the salary that the individual could earn if he or she chose to work for income) is typically used to value losses to the family.[8]

Overall, the loss in national well-being because of environmental degradation includes but is not necessarily limited to the following:

- Loss of healthy life and well-being of the population (premature death and suffering caused by illness)
- Economic losses (such as a decline in soil productivity, the value of other natural resources, or national income, as a result of lost tourism)
- Loss of environmental opportunities (such as the reduced recreational value of lakes, rivers, beaches, and forests).

The process of placing a monetary value on the consequences of environmental degradation typically involves a three step process:

- Quantifying environmental degradation (through monitoring of ambient air quality, water quality, and soil pollution, for example)
- Quantifying the consequences of degradation (such as declines in soil productivity, forest density and growth, natural resource–based recreational activities, and demand for tourism and the impact of by air pollution on health)
- Assigning a monetary value to these consequences.

Environmental and natural resource science, toxicology and epidemiology, economics, and other approaches are used to quantify environmental degradation and its consequences. Economic methods are used to value the consequences of degradation. Where no information on the consequences of degradation is available, primary research is conducted and expert opinions used to develop estimates of likely impacts.

Several approaches can be used to provide quantitative estimates of the consequences of environmental degradation. One method involves computing disability-adjusted life years (DALYs). The main value of this approach, which has been widely used by the World Bank and other institutions, is that it provides a common metric against which to compare impacts on morbidity and mortality. Illnesses are weighted by severity, so that a relatively mild illness or disability represents a small fraction of a DALY while a severe illness represents a larger fraction. One year lost to premature mortality represents one DALY; future years lost are discounted at a fixed rate, typically 3 percent a year.

For waterborne illnesses associated with inadequate water and sanitation services and poor hygiene, the loss of DALYs is caused predominantly by child mortality and population morbidity caused by diarrheal illnesses. Typically, the death of each child under five represents the loss of 35 DALYs.

For outdoor air pollution, impacts on health are estimated based on ambient air quality data from local areas and on local or international studies on adverse health impacts associated with air pollution. Typically, each premature adult death caused by air pollution represents 10 DALYs, based on age-specific cardiopulmonary death rates. Indoor air pollution, particularly in rural areas, can present higher risks than outdoor urban pollution, because of the poor ventilation associated with the use of biomass fuels for cooking and heating. Data for most countries are insufficient to develop estimates of potential health impacts associated with inadequate collection and management of solid waste. In some cases the social costs of inadequate waste management have been estimated directly using willingness-to-pay methods.

The categories of natural resource degradation most commonly quantified include agricultural land and coastal zone degradation, damage from quarries and unplanned construction, deforestation, and disruption of wildlife. For agricultural land degradation, the decline in productivity of land is estimated. The cost of coastal zone degradation is based largely on estimated losses in recreational opportunities, tourism, and ecological values. The impacts of wastewater pollution and inadequate industrial waste management on recreational activities, the value of coastal zones, potable water, and well-functioning ecosystems are typically quantified in more-limited ways. Losses to fisheries are usually not estimated.

A variety of methodologies is applied to monetize the consequences of environmental degradation. In the absence of willingness-to-pay studies, the cost-of-illness approach is usually used for morbidity. This approach estimates treatment costs and the cost of lost work days or time provided by caregivers.[9] The cost of adult mortality from air pollution is estimated based on the willingness to pay for mortality risk reduction. As such studies are not typically available in developing countries, willingness-to-pay estimates from Europe and North America are often used after adjusting for differences in per capita GDP.[10, 11] As a lower bound, DALYs lost to mortality have been valued on the basis of GDP per capita. This valuation technique is similar to the human capital approach, which estimates the cost of mortality as lost future income from the time of death.[12]

A group of 10 assessments sponsored by the World Bank over the past decade is used to examine the application of these methodologies and to compare the results across countries (table 4.3). The earliest assessment—an internal discussion paper completed in 1995—valued environmental costs in Pakistan. *Clear Water, Blue Skies* (World Bank 1998), considered environmental degradation costs as well as some mitigation opportunities in China. Seven of the assessments were conducted under the sponsorship of the Bank's Mediterranean Environmental Technical Assistance Program. They estimate environmental damages in Algeria, the Arab Republic of Egypt, the Islamic Republic of Iran, Lebanon, Morocco, the Syrian Arab Republic, and Tunisia. A study conducted under the sponsorship of

TABLE 4.3
Measured Costs of Environmental Degradation in Selected Countries
(percent of GDP)

Item	Algeria	China	Colombia	Egypt, Arab Rep. of	Iran, Islamic Rep. of	Lebanon	Morocco	Pakistan	Syrian Arab Rep.	Tunisia
Year	1999	1995	2002	2000	2002	2001	1001	1992	2001	2000
Air pollution	1.0	7.1	1.0	2.1	1.6	1.0	1.0	0.9	1.3	0.6
Lack of access to water supply and water sanitation	0.8	0.6	1.0	1.0	2.8	1.1	1.2	2.6	0.9	0.6
Land degradation	1.2	n.a.	0.8	1.2	2.5	0.6	0.4	1.4 (including rangeland and deforestation)	1.0	0.5
Coastal zone degradation	0.6	n.a.	n.a.	0.3	0.2 (Caspian Sea only)	0.7	0.5	n.a.	0.1	0.3
Waste management	0.1	n.a.	n.a.	0.2	0.4	0.1	0.5	n.a.	0.1	0.1
Tourism	n.a.	n.a.	n.a.	n.a.	n.a.	n.a.	n.a.	0.1	n.a.	n.a.
Natural disasters	n.a.	n.a.	0.9	n.a.	n.a.	n.a.	n.a.	n.a.	n.a.	n.a.
Road accidents	n.a.	n.a.	1.5	n.a.	n.a.	n.a.	n.a.	n.a.	n.a.	n.a.
Total measured costs	3.6	7.7	5.2	4.8	7.5	3.4	3.7	5.0	3.3	2.1

Source: Data for Algeria and Morocco are from Sarraf (2004); data for China are from World Bank (1997); data for Colombia are from Larsen (2004); data for the Arab Republic of Egypt are from Sarraf and Larsen (2002); data for the Islamic Republic of Iran are from Sarraf and others (2005); data for Lebanon and Tunisia are from Sarraf, Larsen, and Owaygen (2004); data for Syrian Arab Republic are from Sarraf (2004) and Sarraf, Bolt, and Larsen (2004).
n.a. = Not available.

the Latin American and Caribbean Section of the Bank's Environment Department focuses on Colombia (World Bank 1998; Larsen 2004; Sarraf, Bolt, and Larsen 2004; Sarraf and Larsen 2002; Sarraf, Larsen, and Owaygen 2004; Sarraf and others 2005). Although there are some important methodological differences across studies, all reflect a standard economic paradigm. While some Bank-sponsored studies also report results based on the human capital approach, the estimates considered here are based largely on the willingness-to-pay approach. Some of the methods and, particularly, the underlying dose-response functions, as well as the valuations, have been updated on the basis of new research results. For example, most of the estimates of air pollution damages were based on dose-response coefficients from the international literature on acute effects. Estimates of long-term effects by Pope and others (2002), however, were much larger than previously found. Following extensive peer review, this study is increasingly becoming the convention for estimating mortality effects of particulate pollution. It was used in the two most recent World Bank damage assessments (on Colombia and the Islamic Republic of Iran). It is likely, therefore, that the damage assessments presented in the earlier assessments underestimate the true health damages associated with air pollution. Overall, the methodological harmonization is greatest for air and water pollution. For land and coastal degradation, waste management, and natural resource degradation, the methodologies used depended on the available data.

Clear Water, Blue Skies values a broad array of air pollution damages. In addition to the standard air pollution categories of premature death, morbidity, restricted activity days, chronic bronchitis and other health effects, the authors of the China assessment estimated the cost of lead exposure, mostly from gasoline and stationary sources, such as smelters. They also estimated the effect of acid rain on crops, forests, materials, and ecosystems. Damage by these two sources accounted for about 17 percent of total air pollution damage estimated for China.

The Colombian study (Larsen 2004) includes two damage categories not typically covered in economic/environmental damage assessments: natural disasters and road accidents. Based on data collected from Colombia's Ministry of Interior and Justice, the study estimates that natural disasters such as floods, landslides, avalanches, storms, and earthquakes affected an average of 440,000 people a year over the period 1997–2003, causing damage equivalent to 0.9 percent of GDP.

While road accidents are typically not considered an environmental problem, the World Health Organization (WHO) does include them in its broader definition of the human environment. Accidents were included in the Colombia study for this reason and to provide a comparison with "traditional" environmental issues. The mean annual cost of road accidents is estimated at 1.5 percent of GDP.

Data and resource limitations prevented estimation of degradation costs at the national level for certain damage categories in some countries. All the studies address (outdoor urban) air pollution and water pollution; in contrast, coverage of indoor air pollution, coastal zones (fisheries and tourism losses), wetlands, noise pollution, biodiversity, protected areas, rangeland, inappropriate solid waste disposal, and inadequate industrial and hospital waste management is spottier. The total damages associated with air pollution; water, sanitation and hygiene; and land degradation ranged from 1.7 to 7.7 percent of GDP. Damage caused by water, sanitation, and hygiene represented the largest single category in four countries (Islamic Republic of Iran, Lebanon, Morocco, and Pakistan). Damage caused by air pollution was the most important category in three countries (China, the Arab Republic of Egypt, and the Syrian Arab Republic); in one country (Tunisia) damage caused by air and water pollution was about equal. Land degradation was the most important source of environmental damage only in Algeria.

Degradation of coastal zones, waste management, and tourism accounted for a much smaller share of damage. In the Syrian Arab Republic, for example, these categories accounted for an additional 0.1 percent of lost GDP. An exception is Morocco, where these categories accounted for 1.1 percent of GDP.

The only country that calculated the cost of natural disasters and road accidents was Colombia. The two categories account for 2.4 percent of lost GDP, almost doubling the estimated cost of air pollution; inadequate water, sanitation, and hygiene; and land degradation.

Research shows that both indoor and outdoor air pollution have significant negative effects on human health, causing premature mortality, chronic bronchitis, respiratory disorders, and other effects (table 4.4). The most significant air pollutant in terms of health impacts is particulate matter, especially fine particles (PM_{10} and smaller). The use of biomass fuel for cooking and heating causes indoor air pollution that threatens health. The threat is greatest for women and children, who spend more time indoors than men.

TABLE 4.4
Damage from Indoor and Outdoor Pollution in Selected Countries
(percent of GDP)

Country	Outdoor Air Pollution	Indoor Air Pollution	Total Air Pollution
Algeria	0.6	0.4	1.0
China	5.4	1.7	7.1
Colombia	0.8	0.2	1.0
Egypt, Arab Rep. of	1.8	0.3	2.1
Iran, Islamic Rep. of	1.3	0.3	1.6
Lebanon	0.8	0.2	1.0
Morocco	0.6	0.4	1.0
Tunisia	0.4	0.2	0.6

Source: Data for Algeria and Morocco are from Sarraf (2004); data for China are from World Bank (1998); data for Colombia are from Larsen (2004); data for the Arab Republic of Egypt are from Sarraf and Larsen (2002); data for the Islamic Republic of Iran are from Sarraf and others (2005); data for Lebanon and Tunisia are from Sarraf, Larsen, and Owaygen (2004); data for the Syrian Arab Republic are from Sarraf (2004) and Sarraf, Bolt, and Larsen (2004).

The importance of indoor versus outdoor air pollution varies considerably across the countries studied. In Algeria and Morocco, where total air pollution damage is relatively modest, damage from indoor air pollution is about two-thirds as great as damage from outdoor pollution. In countries with populous and heavily polluted urban centers, the damage from indoor air pollution is relatively smaller, partly because of the smaller number of people using biomass fuels and partly because of the greater health damage caused in urban areas.

On average the effect of indoor air pollution in the countries studied was more modest than in many developing countries in Asia and Sub-Saharan Africa. Indoor air pollution tends to be greater in countries with large rural populations and large shares of households using solid fuels.

Lack of access to clean water ranks as the single most important damage category among the 10 countries studied. Substandard water quality for drinking and hygiene combined with inadequate sanitation facilities led to increases in diarrhea and other waterborne illnesses and associated mortality. Diarrhea and certain other diseases have their greatest impact on children. The economic estimates take account of mortality, morbidity, and the time spent by caregivers taking care of ill people. Because they do not include damage to fisheries, ecosystems, or biodiversity caused by water pollution, these figures underestimate total damage.

In many of the countries studied, soil salinity, erosion, and degradation of rangeland reduce agricultural productivity and the supply of livestock fodder. Although precise data are not available for each source of degradation, estimates of the approximate magnitudes involved are available. In Colombia soil erosion (which reduces crop output) represents about 60 percent of total damage and soil salinity 40 percent (Larsen 2004). (The data did not allow an estimate of the cost rangeland degradation to be made.) In Lebanon, uncontrolled quarrying in the past has caused major destruction of natural vegetation and habitat. Land prices near quarries were significantly lower than for comparable plots farther away (Sarraf, Larsen, and Owaygen 2004). In the Syrian Arab Republic, more than 40 percent of total irrigated land in the Euphrates basin is affected by salinity, which is estimated to reduce yields of cotton and wheat, the main irrigated crops, by 37 percent (Sarraf, Bolt, and Larsen 2004).

Most of the countries studied are located near a major water body.[13] Coastal resources represent an important economic, recreational, and ecological asset. This asset has been threatened by uncontrolled urban development, untreated industrial and municipal discharges, and port activities, which have contributed to significant coastal pollution. For the countries in the Middle East and North Africa, estimates of the cost associated with this pollution include loss of both international and domestic tourism and ecological damage, such as the extinction of sea turtles near the Lebanese coast. In Tunisia, where 90 percent of tourism revenue is associated with coastal zone recreation, a contingent valuation survey found that 12 percent of tourists interviewed were willing to pay about 5 percent more per stay to improve the cleanliness of the beaches.(Sarraf, Larsen, and Owaygen 2004).

Although the 10 studies used different methodologies and covered different categories of damage categories, the results send a powerful message: the economic costs of environmental degradation are significant in virtually every country studied, and they are high in a number of countries. Some of the lower damage estimates reflect dose-response coefficients drawn from earlier studies or the limited availability of data on damage.

A separate issue, not explored in this chapter, concerns intracountry variation in environmental damage. The World Bank analysis of Colombia finds that the distribution of environmental damage varies widely within the country. If documented by other studies, a finding of significant intracountry variation in damages may indicate that policy priorities need to be examined at the subnational level rather for the country as a whole. It is hoped that next generation of studies will take up that challenge.

Conclusions

CRA and economic damage assessment share many similarities. Both identify air and water pollution as key environmental problems and household, industrial, and hazardous waste issues as less significant. The largest differences between the two methods lie in the definitional boundaries of the problems studied. CRAs focus almost exclusively on "brown" issues traditionally managed by pollution control agencies, such as specific air pollutants, drinking water, toxics, and pesticides. They also tend to define problem boundaries on a program-specific basis.

In contrast, economic damage assessments tend to focus on fewer but more broadly defined problem categories, which often cut across specific program lines. They also tend to give more emphasis to "green" issues, such as forestry, agriculture, and coastal management.

There is nothing inherent in the methodologies that dictates these differences; they likely reflect the orientation of the client agencies. For CRAs, the client agencies have generally been involved in environmental management per se. For economic damage assessments, the client agency often includes the local finance and planning ministries or international development agencies, which tend to have a broader perspective on the country's environmental resources.

What circumstances suggest that one approach rather than the other should be applied? Particularly as it has evolved in recent years, the use of CRAs has become participatory in nature, often involving government officials as well as representatives of industry, NGOs, and the general public. In contrast, economic damage assessments are typically conducted by experts with limited involvement in the local political process. While there is nothing inherent in the methodologies that creates these differences, the more-limited public knowledge about and acceptance of environmental economics may be a factor.

How are the results of CRAs or economic damage assessments used in the SEA process? In terms of policy prescriptions, both approaches produce a "worst" list, but this focus on top-ranked problems can be misleading. If the top-ranked risk is global warming, should a country address that issue to the exclusion of all others? While few would advocate such a strategy, particularly in light of the global nature of the problem (and solution), the question highlights one of the dilemmas policymakers face.

Because only a limited number of remedial actions can be costed out, the effectiveness of specific steps to mitigate environmental damages is often unknown. Thus, despite the value of both CRA and economic damage assessments for establishing priority areas for action, they are often insufficient for identifying and evaluating specific projects to mitigate the problems. Doing so involves analyzing marginal benefits and costs of individual actions in order to determine which are likely to provide the largest gain for the resources expended.

Notwithstanding these limitations, both CRA and economic damage assessment can make important contributions to the strategic assessment process. Both provide useful mechanisms for ranking the social impacts of various forms of environmental degradation; both can help policy makers integrate the environment into decisions about economic development.

While economic methods also introduce greater complexities, by casting the issues in clear economic terms, they may make it easier for decision makers to integrate environmental concerns into the broader policy process. These methods give finance and economic ministries a tool for discussing the importance of environmental protections in economic terms, the same "language" used by other ministries.

Some evidence suggests that the results of these studies are being given increasing weight in policy decisions. Keane and Cho (2000) report that CRAs have figured prominently in decisions in Asia, Eastern Europe, and Latin America. Sarraf (2004) reports on the growing use of economic damage assessments in decision making in the Middle East and North Africa. The World Bank and the government of Mexico document the linkage between environmental factors and growth of the tourism sector (World Bank 2005). Buttressed by such analyses, the Mexican government is seeking Bank support to improve infrastructure in and around major tourist facilities.

Particularly as environmental issues continue to enter the mainstream policy arena, a number of opportunities exist to expand the use of these techniques in both national and international decision making. These options may be labeled as carrots and sticks. In the sticks category, the most direct approach is to mandate use of CRA, economic damage assessments, or related techniques. Such mandates could be imposed at the national or subnational level. Another option is for international lending agencies to condition some types of support on the use of such assessments.

To encourage stakeholders to conduct environmental risk and damage studies, international agencies, national governments, or subnational governments could set aside funds for such studies. They could also tailor the scope and complexity of the required analyses better to local conditions and involve more local participants in the studies. Such efforts would not only enhance local capacity-building efforts, they would also help develop a cadre of individuals and institutions better prepared to advance the use of analytical/priority-setting techniques in local decision making.

The increasing clamor for greater efficiency in government programs of all types and the growing emphasis on transparency in government bode well for the expansion of CRA, economic damage assessments, and related techniques in environmental management. To paraphrase Winston Churchill, such techniques may be the worst approach to policy making—except for all the others that have been tried. Indeed, the expanded use of these techniques in environmental management is all but inevitable.

Notes

1 The development of sound management strategies also requires the integration of mitigation cost estimates.

2 Although a growing number of developing countries and international organizations conduct economic damage assessments, this review is confined to recent World Bank studies.

3 For six case studies on the conduct and use of economic analyses of environmental regulations in Europe, see Harrington, Morgenstern, and Sterne (2004).

4 A critique of CRA is that it focuses too narrowly on so-called "brown" issues, at the expense of forestry, fisheries, and other national resource areas. This emphasis reflects the nature of the sponsoring agencies, however, rather than the underlying methodology.

5 Portions of this section are drawn from Morgenstern, Shih, and Sessions (2000).

6 Many studies in OECD countries have assessed the risks associated with environmental problems, and many have tried to compare risks across several environmental issues. A particularly sophisticated series of studies compares the risks associated with different energy technologies (Hirschberg, Spiekernan, and Dones 1998). No CRAs—defined as comparative assessments of most or all of the environmental problems facing a geographic area, the use of risk as the metric for comparison, and priority setting as the fundamental purpose of the work—appear to have been conducted in OECD countries outside the United States, however (for related work, see Navrud and Ready 2002).

7 In some cases the results have been transformed to make them compatible with this reporting format. For example, both *Unfinished Business* and the Taiwan (China) study developed separate rankings for cancer and noncancer health risks. The riskiest individual category was chosen as the overall risk category to combine them into a single health risk ranking. Thus, gas stations, which were ranked as high in terms of cancer risk but medium for noncancer risks, were given an overall ranking of high.

8 For a fuller discussion of these issues, in Sarraf and others (2002, 2004) and Arrow and others (1996).

9 Although not a universally accepted practice, DALYs lost to morbidity are sometimes valued in relation to GDP per capita to estimate the upper value of damages. Such calculations account for the cost of pain and suffering of illness, which is not included in the cost-of-illness approach.

10 The authors of *Clear Water, Blue Skies* (World Bank 1998), which developed estimates of environmental damages in China, scaled a U.S.–based estimate of $3 million per statistical life to a level applicable to China by multiplying the ratio of average per capita incomes in the two countries ($500/$25,000). The resulting estimate was $60,000 per statistical life in urban areas and $31,800 in rural areas of China.

11 Because the elderly (and the very young) are at greatest risk from air pollution, some studies have attempted to adjust willingness-to-pay estimates for differences in life years lost by mortality from air pollution and the overall mortality risks for which the willingness-to-pay estimate was originally calculated. For a discussion of the issue, including a rationale for not making such adjustments, see Alberini and others (2004).

12 The human capital approach—which typically yields lower estimates than the willingness-to-pay approach—has been applied to child mortality by estimating the present value of lifetime income, approximated by GDP per capita, typically for income for people age 20–65.

13 Coastal zone degradation was not studied in China, Colombia, or Pakistan.

References

Alberini, Ana, Maureen Cropper, Alan Krupnick, and Nathalie B. Simon. 2004. "Does the Value of a Statistical Life Vary with Age and Health Status? Evidence from the U.S. and Canada." *Journal of Environmental Economics and Management* 48 (1): 769–92.

Arrow, Kenneth, Maureen L. Cropper, George C. Eads, Robert W. Hahn, Lester B. Lave, Roger G. Noll, Paul R. Portney, Milton Russell, Richard Schmalensee, V. Kerry Smith, and Robert N. Stavins. 1996. *Benefit-Cost Analysis in Environmental, Health, and Safety Regulation: A Statement of Principles.* American Enterprise Institute, the Annapolis Center, and Resources for the Future, Washington, DC.

Blackman, Allen, Richard D. Morgenstern, Libardo Montealegre Murcia, and Juan Carlos Garcia de Brigard. 2005. *Review of the Efficiency and Effectiveness of Colombia's Environmental Policies.* Resources for the Future, Washington, DC.

Feldman, D. L., P. Ralph, and A. H. Rulth. 1996. "Environmental Priority-Setting in U.S. States and Communities: A Comparative Analysis." Energy Environment and Resource Center, University of Tennessee, Knoxville.

Ghosh, Bose & Associates. 1997. "Environmental Management Plan of Asansol-Durgapur Industrial Corridor." Department of Environment, Government of West Bengal, and West Bengal Pollution Control Board. Report prepared with support from the Environmental Health Project and the U.S. Agency for International Development, Regional Housing and Urban Development Office, West Bengal, India.

Harrington, Winston, Richard Morgenstern, and Thomas Sterner, eds. 2004. *Choosing Environmental Policy: Comparing Instruments and Outcomes in the United States and Europe.* Washington, DC: Resources for the Future Press.

Hirschberg, S., G. Spiekernan, and R. Dones. 1998. "Project GaBE: Comprehensive Assessment of Energy Systems: Severe Accidents in the Energy Sector." PSI Bericht Nr. 98–16, Vienna.

ICLEI (International Council for Local Environmental Initiatives). 1995. "Risk-Based Planning: The Troyan Environmental Action Project." Case Study 29, Participatory Priority Setting. Report prepared for the United Nations and the Commission on Sustainable Development, Toronto, April.

ISC (Institute for Sustainable Communities). 1997. "Polish National Environmental Action Program Pilot Project." Report prepared for the U.S. Environmental Protection Agency, the Charles Stewart Mott Foundation, and the German Marshall Fund. Montpelier, VT and the municipalities of Radom and Elk, Poland, Montpelier, VT.

Keane, Susan E., and Jeannie Cho. 2000. "Comparative Risk Assessment in Developing Countries." *Pollution Management in Focus* (January), World Bank, Washington, DC.

Konisky, D. M. 1999. "Comparative Risk Projects: A Methodology for Cross-Project Analysis of Human Health Risk Rankings." Discussion Paper 99–46, Resources for the Future, Washington, DC.

Larsen, Bjorn. 2004. "Cost of Environmental Damage: A Socio-Economic and Environmental Health Risk Assessment." Final report to the Ministry of Environment, Housing and Land Development, Republic of Colombia.

Leonard, Herman B., and Richard J. Zeckhauser. 1986. "Cost-Benefit Analysis Applied to Risks: Its Philosophy and Legitimacy." In *Values at Risk*, ed. D. MacLean. Totowa, NJ: Rowman and Allanheld.

Morgenstern, Richard D., ed. 1997. *Economic Analyses at EPA: Assessing Regulatory Impact.* Washington, DC: Resources for the Future.

Morgenstern, Richard D., and Stuart L. Sessions. 1988. "EPA's Unfinished Business: Weighing Environmental Risks." *Environment* 30 (6): 14–17, 34–39.

Morgenstern, Richard D., Jhih-Shyang Shih, and Stuart L. Sessions. 2000. "Comparative Risk Assessment: An International Comparison of Methodologies and Results." *Journal of Hazardous Materials* 78: 19–39.

Navrud, Ståle, and Richard C. Ready. 2003. "Valuing Cultural Heritage: Applying Environmental Valuation Techniques to Historic Buildings, Monuments and Artifacts." *Journal of Cultural Economics* 27 (3–4): 287–90.

Pope, C. Arden III, Richard T. Burnett, Michael J. Thun, Eugenia E. Calle, Daniel Krewski, Kazuhiko Ito, and George D. Thurston. 2002. "Cardiopulmonary Mortality, and Long-Term Exposure to Fine Particulate Air Pollution." *Journal of the American Medical Association* 287 (5): 1132–41.

Sarraf, Maria. 2004. "Assessing the Costs of Environmental Degradation in the Middle East and North Africa Region." Environment Strategy Note 9, World Bank, Washington, DC.

Sarraf, Maria, Katharine Bolt, and Bjorn Larsen. 2004. *Syrian Arab Republic Cost Assessment of Environmental Degradation.* World Bank/Mediterranean Environmental Technical Assistance Program (METAP), February 9, Washington, DC.

Sarraf, Maria, and Bjorn Larsen. 2002. *Arab Republic of Egypt: Cost Assessment of Environmental Degradation.* Sector Note 25175–EGT, World Bank/Mediterranean Environmental Technical Assistance Program (METAP), Washington, DC.

Sarraf, Maria, Bjorn Larsen, and Marwan Owaygen. 2004. *Cost of Environmental Degradation: The Case of Lebanon and Tunisia.* Environment Department Paper 97, World Bank/Mediterranean Environment Technical Assistance Program (METAP), Washington, DC.

Sarraf, Maria, Marwan Owaygen, Giovanni Ruta, and Lelia Croitoru. 2005. *Islamic Republic of Iran: Cost Assessment of Environmental Degradation.* Report 32043–IR, World Bank/Mediterranean Environmental Technical Assistance Program (METAP), Washington, DC.

Sessions, S., A. Zucchetti, M. Alegre, A. Lanao, and L. Benson. 1997. "Proyecto Ecoriesgo: Ranking Environmental Health Risks in Metropolitan Lima, Peru." Report ANE-0178-Q-00-1047-00, prepared for the U.S. Agency for International Development/Peru by Project in Development and the Environment, Washington, DC.

Slovic, Paul. 1987. "Perception of Risk." *Science* 236 (April 17): 280–85.

Smith, K. R. 1988. "The Risk Transition." Working Paper 10, Environment and Policy Institute, East-West Center, Honolulu, Hawaii.

USAID (U.S. Agency for International Development. 1990. "Ranking Environmental Health Risks in Bangkok, Thailand." Report prepared for USAID Office of Housing and Urban Programs under contract PDC–1008–I–00–9066–00.

———. 1993. "Environmental Health Assessment: A Case Study Conducted in the City of Quito and the County of Pedro Moncayo, Pichincha Province, Ecuador." WASH Reprint: Field Report 436, Environmental Health Division, Office of Nutrition and Health. Washington, DC.

———. 1994. "Comparing Environmental Health Risks in Cairo, Egypt." USAID Project 398–0365, Washington, DC.

———.1995. "A Report on Comparative Environmental Risk Assessment of Ahmedabad City." Report prepared under contract 386-1008-0-00-4286 by the Centre for Environmental Planning and Technology for the USAID Regional Housing and Urban Development Office, Washington, DC.

———. 1996. "Comparative Risk Assessment for Central America: Executive Summary." Report ANE-0178-Q-00-1047–00, prepared for USAID/Guatemala by Project in Development and the Environment and Chemonics International, Washington, DC.

U.S. EPA (United States Environmental Protection Agency). 1987. *Unfinished Business: A Comparative Assessment of Environmental Problems.* Office of Policy Analysis, Washington, DC.

———. 1990b. *Reducing Risk: Setting Priorities and Strategies for Environmental Protection.* Science Advisory Board, Washington, DC.

———. 1992. "Project Silesia: Comparative Risk Screening Analysis." Report prepared for the Technical Workgroup, Ostrava (former Czechoslovakia) and the U.S. Environmental Protection Agency by Industrial Economics, Inc., and Sullivan Environmental Consulting, Washington, DC.

———. 1994. "Project Silesia: Comparative Risk Screening Analysis." Report prepared for the Technical Workgroup, Katowice, Poland, the U.S. Environmental Protection Agency by Industrial Economics, Inc., and Sullivan Environmental Consulting, Washington, DC.

———. 1997. *The Benefits and Costs of the Clean Air Act: 1970 to 1990.* Office of Air and Radiation/Office of Policy, Washington, DC.

World Bank. 2005. "Evaluacion ambiental estrategica del sector turismo en Mexico." Study published in coordination with Environmental Resources Management and the Ministry of Tourism, Mexico, June 30.

Bibliography

Becker, H. 1997. *Social Impact Assessment.* London: University College Press.

Baiocchi, Gianpaolo, Patrick Heller, Shubham Chaudhuri, and Marcelo Kunrath Silva. 2005. "Evaluating Empowerment: Participatory Budgeting in Brazilian Municipalities." World Bank Policy Research Working Paper, Washington, DC.

Beddies, S., and H. De Soto. 2005. *Poverty and Social Impact Analysis (PSIA) of the Decentralization and Water Sector Privatization in Albania*. World Bank, Social Development Department, Washington, DC.

Bianchi, R., and S. Kossoudji. 2001. *Interest Groups and Organizations as Stakeholders*. Social Development Paper 35, World Bank, Washington, DC.

Blackburn, James, Robert Chambers, and John Gaventa. "Mainstreaming Participation in Development." World Bank, Operations Evaluation Department, Washington, DC.

Brinkerhoff, D., and B. L. Crosby. 2002. *Managing Policy Reform: Concepts and Tools for Decision-Makers in Developing and Transition Countries*. Bloomfield, CT: Kumarian Press.

Chambers, R. 1983. *Putting the First Last*. London: Longman.

Dedu, G., and G. Kajubi. 2005. "The Community Scorecard Process in The Gambia." Social Development Note 100, World Bank, Washington, DC.

DeWind, Josh, and David H. Kinley. 1988. *Aiding Migration: The Impact of International Development Assistance on Haiti*. Boulder, CO: Westview Press.

Dulamdary, E., M. Shah, and R. Mearns, with B. Enkhbat and L. Ganzaya. 2001. "Mongolia: Participatory Living Standards Assessment." Report prepared for the Donors Consultative Group Meeting, Paris, May 15–16. National Statistics Office, Ulaanbaatar, Mongolia, and World Bank, Washington, DC.

Feldman, M., and A. Khademian. 2000. "Management for Inclusion: Balancing Control with Participation." *International Public Management Journal* 3 (2): 149–68.

———. 2005. "Inclusive Management: Building Relationships with the Public." School for Public and International Affairs, Virginia Tech, Blacksburg, VA.

Finsterbusch, K., J. Ingersoll, and L. Llewellyn. 1990. *Methods for Social Analysis in Developing Countries*. San Francisco: Westview Press.

Goldman, L. R., ed. 2000. *Social Impact Analysis: An Applied Anthropology Manual*. Oxford: Berg Press.

Government of Mongolia. 2003. *Economic Growth Support and Poverty Reduction Strategy*. Ulaanbaatar.

Grindle, M. S., and J. W. Thomas. 1991. *Public Choices and Policy Change: The Political Economy of Reform in Developing Countries*. Baltimore, MD: Johns Hopkins University Press.

Haney, M., M. Shkaratan, V. Kabalina, V. Paniotto, and C. Rughinis. 2003. *Mine Closure and Its Impact on the Community: Five Years after Mine Closure in Romania, Russia and Ukraine*. Social Development Paper 42, World Bank, Washington, DC.

Holland, Jeremy, and James Blackburn. 1998. *Whose Voice? Participatory Research and Policy Change*. Intermediate Technology Development Group, London.

Junge N., T. Pushak, J. Lampietti, N. Dudwick, and K. Van den Berg. 2004. *Sharing Power: Lessons Learned from the Reform and Privatization of Moldova's Electricity Sector*. World Bank Policy Research Working Paper, Washington, DC.

Kanji, N., and S. Ware. 2002. "Trade Liberalization, Poverty and Livelihoods: Understanding the Linkages." Department for International Development, African Policy and Economic Department, London.

Keener, S., and S. Banerjee. 2006. "Ghana: Electricity Tariff Reform." In *Poverty and Social Analysis*, ed. A. Dani. Washington, DC: World Bank.

Kvam, R., and H. Nordang. "The Jharkand Participatory Forest Management Project: Social Assessment for Inclusion, Cohesion, and Accountability." World Bank, Washington, DC.

Leitner, Kerstin. 2005. "Heath and Environment: A View from WHO." *In Environment Matters.* Washington, DC: World Bank.

Marquetti, Adalmir. 2000. "Participatory Budgeting in Porto Alegre." *Indicator S A* (Johannesburg) 17 (4): 71–78.

Mearns, R. 2004. "Sustaining Livelihoods on Mongolia's Pastoral Commons: Insights from a Participatory Poverty Assessment." *Development and Change* 35 (1): 107–39.

Menegat, Rualdo. 2002. "Participatory Democracy and Sustainable Development: Integrated Urban Environmental Management in Porto Alegre, Brazil." *Environment and Urbanization* 14 (2): 181–206.

Menegat, Rualdo, Maria Luiza Porto, Clovis Carlos Carrazo, and Luís Alberto Dávila Fernandes. 1998. *Environmental Atlas of Porto Alegre.* Porto Alegre, Brazil: Edufrgs.

Narayan, D., R. Chambers, M. Shah, and P. Petesch. 2001. *Voices of the Poor: Crying Out for Change.* Washington, DC: World Bank.

National Statistics Office of Mongolia, and World Bank. 2001. *Mongolia Participatory Living Standards Assessment 2000.* Ulaanbaatar.

Norton, A., B. Bird, K. Brock, M. Kakande, and C. Turk. 2001. *A Rough Guide to Participatory Poverty Assessments: An Introduction to Theory and Practice.* London: ODI Publications.

Ostro, Bart. 1994. "Estimating the Health Effects of Air Pollutants: A Methodology with Application to Jakarta." Policy Research Working Paper 1301, World Bank, Washington, DC.

Paul, S. 2002. *Holding the State to Account: Citizen Monitoring in Action.* Bangalore: Books for Change.

Pollard, Amy. 2005. *How Civil Society Organisations Use Evidence to Influence Policy Processes: An Annotated Bibliography.* Overseas Development Institute, London.

Public Affairs Center. 2002. *The State of Karnataka's Public Services: Benchmarks for the New Millennium.* Bangalore.

Rawls, John. 1971. *A Theory of Justice.* Cambridge, MA: Belknap Press of Harvard University Press.

Robb, C. 2000a. *Can the Poor Influence Policy? Participatory Poverty Assessments in the Developing World.* Washington, DC: International Monetary Fund and World Bank.

———. 2000b. "How the Poor Can Have a Voice in Government Policy?" *Finance and Development* 37 (4). International Monetary Fund, Washington, DC.

———. 2003. "Poverty and Social Impact Analysis: Linking Macroeconomic Policies to Poverty Outcomes: Summary of Early Experiences." IMF Working Paper WP/03/43, International Monetary Fund, Washington, DC.

Robb, C., and A. Scott. 2001. "Reviewing Some Early Poverty Reduction Strategy Papers in Africa." IMF Policy Discussion Paper PDP/01/5, International Monetary Fund, Washington, DC.

Salmen, L. 1995. "Listening to the People." *Finance and Development* 32 (2): 44–48. International Monetary Fund, Washington, DC.

———. 2002. *Beneficiary Assessment: An Approach Described.* Social Development Paper 10, World Bank, Washington, DC.

Salmen, L., and M. Amelga. 1998. *Implementing Beneficiary Assessment in Education: A Guide for Practitioners.* Social Development Paper 25, World Bank, Washington, DC.

Salmen, L., and E. Kane. 2006. *Bridging Diversity: Participatory Learning for Responsive Development.* Washington, DC: World Bank.

Shah, P., and D. Youssef. 2002. "Voices and Choices at a Macro Level. Participation in Country-Owned Poverty Reduction Strategies." Action Learning Program Dissemination Series 1, World Bank, Social Development Department, Washington, DC.

Subbarao, Kalanidhi, Kene Ezemenari, Aniruddha Bonnerjee, Soniya Carvalho, Alan Thompson, Carol Graham, and Jeanine Braithwaite. 1997. *Safety Net Programs and Poverty Reduction: Lessons from Cross-Country Experience.* Washington, DC: World Bank.

Thindwa, J., C. Monico, and W. Reuben. 2003. "Enabling Environments for Civic Engagement in PRSP Countries." Social Development Note 82, World Bank, Washington, DC.

Tikare, S., D. Youssef, P. Donnelly-Roark, and P. Shah. 2001. "Organizing Participatory Processes in the Poverty Reduction Strategy Process." In *PRSP Sourcebook.* Washington, DC: World Bank.

Turk, C. 2001. "Linking Participatory Poverty Assessments to Policy and Policy Making: Experience from Vietnam." Policy Research Working Paper 2526, World Bank, Washington, DC.

Wagle, S., and P. Shah. 2003. "Porto Alegre, Brazil: Participatory Approaches in Budgeting and Public Expenditure Management." Social Development Note 71, World Bank, Washington, DC.

Wagle, S., J. Singh, and P. Shah. 2004. "Citizen Report Card Surveys: A Note on the Concept and Methodology." Social Development Note 91, World Bank, Washington, DC.

Whittingham, E., J. Campbell, and Philip Townsley. *Poverty and Reefs.* London: Department for International Development.

World Bank. 1998. *Clear Water, Blue Skies.* Washington, DC: World Bank.

———. 2001a. *Filipino Report Card on Pro-Poor Services.* Washington, DC.

———. 2001b. *Making Sustainable Commitments: An Environment Strategy for World Bank.* Washington, DC: World Bank.

———. 2002. *World Development Report 2003: Sustainable Development in a Dynamic World.* Washington, DC: World Bank.

———. 2003a. *Cornerstones for Conservation: World Bank Assistance for Protected Areas.* Washington, DC.

———. 2003b. *Indonesia Kerinci Seblat Integrated Conservation and Development Project: Implementation Completion Report.* Washington, DC.

———. 2003c. *A User's Guide to Poverty and Social Impact Analysis.* Washington, DC: World Bank. http://www.worldbank.org/poverty/psia/index.htm.

———. 2004a. "Community Scorecard Process: A Short Note on the General Methodology for Implementation." World Bank, Washington, DC.

———. 2004b. *Responsible Growth for the New Millennium.* World Bank: Washington, DC.

———. 2005. *Indonesia Coral Reef Rehabilitation and Management Project (COREMAP): Implementation Completion Report.* East Asia and Pacific Region, Washington, DC.

World Bank, and DFID (Department for International Development). 2005. *Tools for Institutional, Political and Social Analysis (TIPS) in Poverty and Social Impact Analysis (PSIA): A Sourcebook for Commissioners and Practitioners,* vol. 1. Washington, DC, and London.

World Health Organization. 2002. *World Health Report.* Geneva: WHO.

Zeckhauser, Richard J. 1981. "Preferred Policies When There Is Concern for Probability of Adoption." *Journal of Environmental Economics and Policy* 8 (3): 215–37.

C H A P T E R   5

Giving the Most
Vulnerable a Voice

Caroline Kende-Robb and
Warren A. Van Wicklin III

IN MANY COUNTRIES, the most vulnerable often have the least influence on policy formulation, even if they are the most affected by policies. This has begun to change as governments and donors have developed tools and opened up policy-formulation processes to enable greater public participation, including that of the most vulnerable. To increase transparency and improve governance, an enabling environment is required that ensures that citizens participate in natural resource management policy making and implementation and contribute to the management of resource allocation.

Participatory approaches have been used at the project level for some time. Their use for policy analysis is more recent. Policy formulation was considered too technical for most people, but as governments and donors adopted participatory processes more generally, they found that people have relevant knowledge and can make useful contributions to policy. This progression to the policy level is a logical extension of giving the most vulnerable increasing voice and participation at the project level.

Caroline Kende-Robb is a sector manager in the Social Development Department at the World Bank. Warren A. Van Wicklin III is an international development consultant. The authors thank Robin Mearns for inputs to the Mongolia case, Andre Herzog for inputs to Brazil example, and Sean Doolan for inputs on environmental governance.

Shifting environmental considerations to the policy level has improved the ability to generate a better range of options from an environmental perspective. Shifting participation to the policy level can have similar benefits. Rather than providing input about a specific dam, for example, citizens can help shape regional or national water and energy strategies, including demand management and alternative sources of water and energy supply. Instead of mitigating impacts through restrictions on the use of natural resources in protected areas, citizens can help formulate policies on common property resources, such as comanagement and sustainable use.

Governance over the environment and natural resources urgently needs improvement. Changes in patterns of natural resource exploitation (including global market and political linkages, international concerns along the value chain, and shifting national institutional and political perspectives) have created social tensions at the local level. Increasing climatic variability exacerbates many of these pressures. The equity and transparency of revenue sharing from natural resources are often highly contested, and conflicts between communities, the private sector, and the state over access to natural resources are escalating. Much of the conflict and depletion of natural resources can be traced to poor management of natural resources, weak environmental protection, and a lack of voice for those most affected. In many countries the sustainability of the natural resource sectors and prospects for community stability and economic growth are put at risk by the absence of effective regulatory institutions, by weak mechanisms for citizen voice, and by indecisive leadership on natural resources and the environment. Building mechanisms and institutions that increase citizen voice can help reduce social tensions and conflicts associated with natural resource and environmental management and revenue allocation.

Studies show that by giving the most vulnerable a voice, policy makers are better able to understand the synergies between environmental goals, economic growth, and poverty reduction (World Bank 2001). The results of consultative and participatory processes highlight that livelihood strategies adopted by vulnerable groups are inextricably linked to the environment: degraded environments increase poverty, while poverty often degrades the environment (World Bank and DFID 2005).

This chapter reviews recent experience on how vulnerable groups can be given greater voice in policy formulation, especially formulation of policies with environmental considerations. It is divided into eight main sections. The first explains why it is important to include vulnerable groups in policy formulation. The second defines people who are vulnerable from an environmental perspective. The third discusses the importance of an enabling environment in creating "space" for participation. The fourth identifies entry points in the policy process for increasing voice. The fifth defines five levels of participation. The sixth describes seven common tools for amplifying the voice of vulnerable groups as well as making sure they are heard in the policy process. The seventh describes case studies in which vulnerable groups had a significant voice in policy formulation. The section draws some preliminary conclusions on the experience to date.

Why the Vulnerable Should Be Involved in Policy Formulation

There are many reasons to give vulnerable groups greater voice, including technical and ethical rationales.[1] They include direct benefits for vulnerable groups (intrinsic value) and benefits for policy formulation (instrumental benefits) (figure 5.1).

First, including vulnerable groups leads to better policy analysis. Vulnerable households are often the first to experience the direct and indirect impacts of policies and to be most affected by them. Only they can truly feel and explain their experiences and perspectives. Experience has shown that vulnerable groups have the capacity to appraise, analyze, plan, act, and monitor to a far greater extent that had previously been acknowledged or assumed (Holland and Blackburn 1998; Robb and Scott 2001; Chambers 2007).

Better-informed technical diagnosis leads to better policies. Including vulnerable groups stimulates debates on policy impacts and trade-offs based on experience.[2] Public debate can help identify the most appropriate policy combination to promote growth, reduce poverty, and protect the environment. Policies formulated by a broader range of stakeholders are likely to have fewer unanticipated and unintended consequences and to be more predictable in their impacts.

Second, participatory processes help foster understanding, ownership, and support of policies and their effective implementation. If previously excluded

FIGURE 5.1
Benefits of Participation and Consultation in Policy Formulation

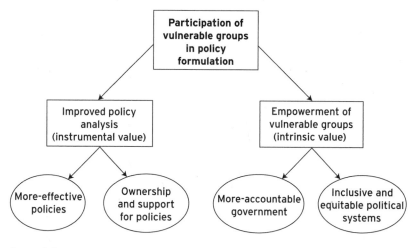

Source: Authors.

vulnerable groups are given a sufficient role in policy formulation, they are more likely to accept the results of those processes, less likely to resist them, and perhaps more likely to support those policies if they have sufficient incentives to do so (Blackburn, Chambers, and Gaventa 2000).

Third, participatory policy formulation helps marginalized and vulnerable groups develop a relationship with government and other stakeholders.[3] More-equitable, more-inclusive political systems are promoted through an appreciation of the varied needs of all groups within society. There is growing recognition that citizens have an important role to play in enhancing the accountability of public officials and the private sector and in reducing corruption and the leakage of funds. Social accountability has become an attractive approach for improving governance processes, service delivery outcomes, and resource allocation decisions (see Ackerman 2005).[4] Over the past decade, numerous examples have emerged that demonstrate how citizens can make their voices heard and make the public and private sectors more responsive and accountable. In many countries the capacity of civil society needs to be strengthened to ensure that citizens understand their rights and obligations, build stronger partnerships, participate in local- and central-level policy-making processes, and monitor the delivery of services of public and private sector institutions.

Fourth, participation contributes to more-accountable government. The accountability of institutions is strengthened; citizens, including vulnerable groups, are better informed about government commitments; and governments are held accountable to their constituencies for performance. Building the role and capacity of nonstate institutions can help counterbalance the power of the state.

Fifth, citizens are empowered by their participation in policy formulation and implementation. By helping shape policies to make them more relevant and responsive to their needs, they no longer feel as vulnerable, insecure, or powerless. They become mobilized and active citizens. Participation in policy formulation often leads to other actions that improve their position, such as developing partnerships with like-minded stakeholders. Empowerment of vulnerable groups and accountability also help prevent elite capture of the policy process.

Last, there is an ethical argument for giving the excluded a voice. It is only fair that those who are likely to be most severely affected have a voice. Policies should aspire to have fair, if not equitable, effects.[5] They should be crafted so that their rationale is independent of any one stakeholder's position and can be justified on behalf of society as a whole.

Definition of the Vulnerable and How They Are Affected

Policy analysis has traditionally focused on a statistical approach to poverty based on indicators of income, health, and education. Poverty itself was measured by income. Studies show that an approach dominated by economic analysis fails to capture many dimensions of poverty and that a multidisciplinary approach can deepen understanding of the poor and vulnerable (Robb 2000).

The World Bank's *Voices of the Poor* study helped elaborate a multidimensional definition of poverty focused on vulnerability, physical and social isolation, insecurity, lack of dignity and self-respect, lack of access to information, distrust of state institutions, and powerlessness to defend one's interests or change things (Narayan and others 2001). These dimensions are often interlinked. Many people are at the intersection of multiple forms of vulnerability. Poverty alone is not sufficient to define vulnerable groups.

The poor are not a static group. Economic transition and shocks throw people into poverty and expose their vulnerability. Former socialist economies created vulnerability as people lost guaranteed employment and resorted to natural resource–based livelihoods that led to environmental degradation. Many Mongolian state employees who lost their jobs, for example, went into goat herding, the nation's leading occupation. The sudden unmanaged increase in livestock herds put unsustainable pressures on pasture land (Mearns 2004). The East Asian financial crisis reversed decades of impressive poverty reduction, throwing many people back into poverty. In the struggle to survive economic shocks, many turned to common property resources to sustain themselves, accelerating deforestation and other forms of environmental degradation (World Bank 2003b).

Maximizing income may be less of a priority to vulnerable groups than securing their livelihoods and reducing their vulnerability (Chambers 1983). For livelihoods to be sustainable, they need to be resilient to shocks without undermining the natural resource base that supports them (Kanji and Ware 2002). Vulnerability appears to be a more salient concept than poverty. It is a broader and more inclusive concept. It may also be more relevant, especially in terms of policies that have environmental consequences.

Some demographic and social groups are vulnerable for reasons other than economic factors. Ethnic minorities and indigenous people, for example, are often vulnerable because of exclusion or domination by the majority ethnic group. Women, children, and the elderly are often vulnerable even when they are not poor. Women, too, often do lack access to or control over resources and discrimination in societies that exclude them from decision making. In the extractive industries sector, for example, most benefits (employment, income, royalties, and infrastructure) tend to accrue to men, whereas negative impacts (cultural disruption, social stress, environmental harm, and domestic violence) affect women and children.

As defined here, people are vulnerable either because they depend on the environment for their livelihoods or because they are otherwise significantly affected by environmental degradation.

Where economic and environmental vulnerability are mutually reinforcing, these factors are often interconnected. Examples include the following:

- Natural resource management (forests, water, land, and other natural resources)
- Environmental health (outdoor and indoor air pollution, waterborne diseases, occupational health and safety, toxins, and other sources of risk)
- Waste management (solid waste, water pollution)
- Vulnerability to natural disasters (short term) and climate change (long term).

Natural Resource Management

Stronger governance in the natural resources and environment sectors is important to ensure that these sectors continue to contribute responsibly to future patterns of inclusive growth. Strengthening governance requires effective regulatory mechanisms and political commitment to manage the environmental resource base and control adverse environmental impacts that affect livelihoods and levels of human development. It also requires more-open access to environmental information and analyses, expressed in terms that are relevant to other stakeholders, accompanied by mechanisms that enable broader dialogue across multiple stakeholders down to the local level. Vulnerability in the context of natural resources occurs when livelihoods are threatened by a change in environmental conditions of natural resources or in access to these resources. The poor are often disproportionately affected by environmental degradation (World Bank 2001).

Globally, 1.3 billion people live on fragile lands—arid zones, slopes, wetlands, and forests—that cannot sustain them (World Bank 2002). The majority of the world's poor and vulnerable depend largely on agriculture for their livelihood. As natural resources become scarce or degraded, their incomes and livelihoods are threatened. Increasing desertification has increased the vulnerability of millions of pastoralists, especially in Africa. As water supplies become less reliable, poor farmers are often the most vulnerable: wealthier farmers can often afford wells and irrigation systems, but poor farmers are dependent on rainfed agriculture.

Sometimes vulnerable groups may be a particular set of stakeholders who have limited influence. About 60 million people in the world (mainly indigenous and tribal groups) depend almost entirely on forests; another 350 million people who live within or adjacent to dense forests depend on them to a high degree for subsistence and income (World Bank 2004b). Of the world's 1 billion extreme poor—those living on less than a dollar a day—90 percent are significantly dependent on forest resources for their income (World Bank 2002). Deforestation or use of these forests for other purposes threatens the livelihoods of these groups. Logging, mining, and commercial agricultural interests are usually more environmentally destructive than the activities of the local poor, who have practiced sustainable forestry, farming, hunting, and gathering for generations. Their lack of political and economic power means their interests are often subverted to wealthier and more powerful interests. Vulnerable groups need greater voice for both economic and environmental reasons. The creation of protected areas, for example, should protect the livelihoods and property rights of people living within those areas (World Bank 2003a).

The causal linkage between economic and environmental vulnerability often works in the other direction as well, with poverty causing environmental degradation and exacerbating vulnerability in a vicious circle. In many poor countries, the best agricultural land is owned by the wealthiest farmers. Many poor farmers have been forced to find farmland in marginal areas, such as steeply sloped land that is unsuitable for mechanized agriculture. Farming steeper slopes, especially for unsuitable food crops, accelerates their erosion and degradation. In Haiti this practice led to massive environmental degradation, impoverishment, and migration (DeWind and Kinley 1988).

The livelihoods of other people who depend on natural resources are also adversely affected by environmental impacts. Artisanal fishers suffer as coastal areas and coral reefs are overfished, become excessively polluted, or are otherwise degraded (Whittingham, Campbell, and Townsley 2003). In Indonesia cyanide and dynamite fishing has destroyed or badly damaged coral reefs that support fish populations. Many of the most destructive fishing practices are engaged in not by the local poor but by wealthier outsiders, who then move on to new fishing areas once an area has been destroyed and fished out. Local fishers, with their need for more sustainable fishing practices, are often the strongest advocates for environmentally sensitive fishing regulations and creation of protected marine areas (World Bank 2005).

Environmental Health

Vulnerability in the context of environmental health occurs when people are susceptible to increased health risks associated with environmental pollution. These risks disproportionately affect the poor and children under five.

Lack of access to clean water and sanitation and indoor air pollution are the two main causes of illness and death in developing countries. Both problems principally affect children under the age of five (as reflected in high infant mortality rates linked with diarrhea, pneumonia, acute respiratory infections, and water-borne diseases [WHO 2002]).

Increased risk of waterborne diseases caused by inadequate water quality, sanitation, and hygiene is also a major cause of illnesses and death in developing countries (WHO 2002). Diarrhea causes 1.7 million deaths a year and accounts for about a third of deaths among children under the age of five in developing countries (WHO 2002). Worldwide an estimated 3 million people a year die from water-related diseases (World Bank 2001). More than 90 percent of health effects are experienced by children under age five.

Indoor air pollution comes from the burning of fuel wood for cooking and heating. Biomass is the primary source of energy for about 2.5 billion people (IEA 2006). It causes acute respiratory infections in children and increased risk of chronic pulmonary disease in women. In developing countries with high mortality rates, indoor air pollution is the fourth leading cause of illness and death (WHO 2002). Nearly 2 million children and women die every year from diseases caused by indoor air pollution (World Bank 2001), most of them from poor families (wealthier households use cleaner fuels).

Substantial evidence from around the world indicates that outdoor urban air pollution has significant negative impacts on public health and results in premature death, chronic bronchitis, and respiratory disorders (Ostro 1994). The burden of disease is lower than for indoor air pollution, however (WHO 2002). Even so, 1 million people a year in developing countries die prematurely because of urban air pollution, a disproportionate number of them children under five (Leitner 2005).

Waste Management

Vulnerability in the context of waste management occurs when people are susceptible to increased health risk. Where the increased health risk comes from toxic chemicals leaching into the soil and affecting water supplies, people living near the waste dumps will be principally affected. Waste pickers are directly affected by vector-borne diseases.

Natural Disasters and Climate Change

Vulnerability in the context of natural disasters occurs when people's livelihoods, assets, and health are adversely affected by natural disasters. Greater climate variability as a result of global climate change is expected to increase the number of severe weather events over time and change the absolute level of certain natural resources, such as water and soil productivity, affecting both health and livelihoods.

The poor are more vulnerable to natural disasters such as mudslides, tsunamis, earthquakes, and floods, for several reasons. They often live in areas that are more prone to such disasters, such as steep hillsides, along flood channels, or in low-lying areas. Their houses are generally less sturdy and thus more easily damaged than those of the nonpoor. They usually have less information about impending catastrophes and cannot avail themselves of warning systems as efficiently. Their economic vulnerability also places them at greater risk.

The poor are also more vulnerable to long-term environmental degradation. The poor are least able to afford mitigating or avoiding the impacts of rising sea levels, and they tend to live at lower elevations, which are more vulnerable to flooding. Some South Pacific islands are already losing land to rising sea levels and relocating their inhabitants. Changing crop patterns are also more likely to have disproportionate effects on the poor, who find it expensive and more difficult to adapt to new crops, new locations, or new agricultural practices. An informed national consituency is needed to drive potential responses to climate change. Ensuring public debate on the impacts of climate change and adaptation to climate change that moves beyond technical perspectives to incorporate local voices would help improve natural resource and environmental governance.

Creation of an Enabling Environment in Which the Vulnerable Can Be Heard

The entry points, forms of participation, and tools for providing voice for the environmentally vulnerable are the same as those for vulnerable groups in general. Local dimensions of environmental challenges often vary across a country. Developing mechanisms for recognizing these issues at the national level and enabling polic makers to make decentralized responses requires an enabling environment in which citizens' voice can be heard at the policy-making level. The next four sections of this chapter therefore draw from the general literature on inclusive decision making. In fact, many of the tools were developed for use at the project level and are therefore not even specific to policy formulation.

In chapter 3 of this volume, Feldman and Khademian present a decision-making model that has two main features: adaptive management and inclusive decision making. Adaptive management is a model of continuous policy formation. The basic idea of the model is to try something and then evaluate the outcome in order to determine what to try next. Policy is formed through a process of experimentation, the adoption of successful experiments and continued experimentation. Involving people and organizations with a wide range of perspectives in a process of policy making and implementation is seen as legitimate and an essential outcome of inclusive management. Where traditional management models may see building community capacity as a by-product of solving policy problems, this model places capacity building in the foreground. Building capacity becomes the primary goal, which is pursued through projects. Combining adaptive management with the creation of an inclusive community of participation and the consequent acknowledgment and exposure of ambiguity create the inclusive model of policy formation (Feldman and Khademian 2000, 2005).

Directly consulting poor people does not ensure that they will have a voice in influencing policy formulation. Policy making is an inherently political process. Rules, legislation, traditions, networks, alliances, patronage, and bureaucratic structures interact to form a complex and fluctuating policy environment (Grindle and Thomas 1991; Robb 2002). The political context and institutional framework through which consultation and participation take place can be built on to promote a more open process of policy formulation. Policy making is often a negotiation between groups of unequal power and influence, in which the poor and vulnerable have the least power. An appreciation of the unequal underlying power relations can lead to a better understanding of how such policy choices are made and how different groups are affected by them.

Inclusive decision making depends on the existence of an enabling environment, which depends in large measure on the extent to which certain external factors assist or hinder promotion of the interests of vulnerable groups. Such factors include the legal and regulatory framework, the political and governance context, sociocultural characteristics, and economic conditions. These external conditions in turn influence specific enabling elements that are essential to the effectiveness of civil society as a key determinant of development. These include the freedom of citizens to associate; their ability to mobilize financial resources to fulfill the objectives of their organizations; their ability to formulate, articulate, and convey opinion; their access to information (necessary for their ability to exercise voice and engage in negotiation); and the existence of spaces and rules of engagement for negotiation and public debate (Thindwa, Monico, and Reuben 2003).

Institutional and organizational dynamics within civil society also influence the enabling elements. Along with the external factors, they must be the subject of analysis if the fullest scope of constraints to giving vulnerable groups voice is to be understood. These dynamics include such factors as accountability, representation, legitimacy, institutional and organizational capacities, self-regulation, and institutional relationships across civil society groups and between civil society groups and the state and private sector.

Entry Points for Giving Vulnerable Groups Voice

According to Feldman and Khademian (chapter 3), policy formulation is an ongoing, iterative, adaptive process; participation at any stage gives voice in the next cycle of policy making, including by affecting ongoing implementation of the existing policy. For this reason, participation needs to be an ongoing process, not a one-off event, such as a consultation or workshop. The entire policy process cycle presents opportunities for participation. In some contexts institutional processes include specific spaces for consultation, negotiation, and debate among different stakeholders, including vulnerable groups. In many other contexts, no such spaces exist. In these settings participation can continuously raise issues, so that over time the needs of vulnerable groups are taken into account through incremental changes in a particular policy.

The policy process is not linear or coherent. In policy development and implementation, change is often multidirectional, fragmented, frequently interrupted, and unpredictable. How to sequence actions, what to pay attention to, and whom to include can be hard to determine, and it can vary over the life of the policy cycle. Unlike projects and programs, which have clearly defined sets of activities and components, the boundaries for policies tend to be fuzzy, to shift over time, and to be open to interpretation. Multiple stakeholder institutional arrangements can help build consensus in key areas of policy formulation and implementation. There is a range of opinion about the structure of policy processes. Moreover, steps overlap and are not always in the same order. All policy processes, however, involve the following four steps (Shah and Youssef 2002):

- *Analysis.* Once a decision has been made to promote a new policy or reform an existing one, a range of analyses and diagnoses is undertaken. Ideally, the first step is to conduct a stakeholder analysis to determine which groups may be affected by the policy or could have an interest in shaping it. Vulnerable groups can be directly consulted to analyze potential impacts and the impacts of past policies. Not only can vulnerable groups help develop the policy; they can also help develop the consultation process.

- *Formulation.* Once the relevant information is available, policy formulation adds operational detail to the initial policy analysis, which includes a statement of the problem, the policy goals and objectives, a framework that sketches programs in support of those targets, and a statement of required resources.

- *Implementation.* Entry points for external participation in implementation vary, depending on the type of policy. Some sectoral policies, for example, offer opportunities for partnerships for service delivery in which external entities may take the lead through contracting-out, delegation of authority, or community comanagement. The vulnerable should not only help develop the policy but have a role in its implementation, to ensure that the policy is implemented as intended and that commitments are honored.

- *Monitoring and evaluation.* The vulnerable can provide information to policy monitors and evaluators, and they can be consulted for their opinions, interpretations, and analyses. They can collaborate in joint monitoring and evaluation, or they can conduct their own independent assessments and evaluations. They can provide valuable data on the poverty and social impacts of policies and on institutional performance. They can help amend policies during implementation if necessary. They feed information into policy-making processes in order to enable policy makers to better understand the potential impacts of future policies and reforms.

Including vulnerable groups at various stages can be considered good-practice policy making because it makes assumptions more explicit up front (including the poor in ex ante analysis); monitors whether public actions and choices are working, thereby testing original assumptions and taking midcourse changes if public actions are not succeeding (including vulnerable groups during policy implementation); assesses whether public actions were successful; and uses the information to influence future policy design (including the poor in ex post evaluation).

Levels of Participation in Policy Processes

Participation ranges from information dissemination and consultation to joint decision making (Robb and Scott 2001) (figure 5.2):

- Information dissemination is the one-way, top-down flow of information to the public about the impacts of ongoing policies and policy diagnostics for the proposed policy or policy change. Knowledge about policies gives vulnerable groups the opportunity to analyze and attempt to influence policies.
- Consultation is the two-way sharing of information. It gives vulnerable groups the chance to share their knowledge and perspectives and to inform policy makers and other stakeholders. The very act of being consulted empowers vulnerable people, validating their knowledge.
- Joint analysis is a process in which vulnerable groups and policy makers join to analyze information relevant to the policy.

FIGURE 5.2
Flowchart of Participatory Policy Processes

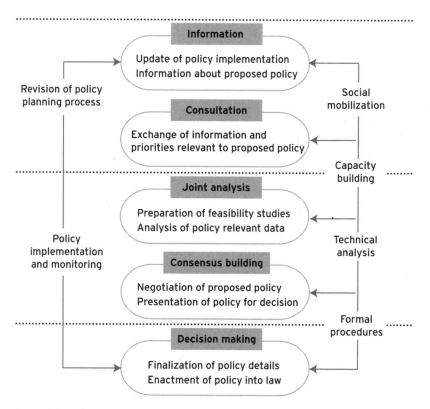

Source: Adapted from Malena, with Forster and Singh 2004.

■ Consensus-building goes beyond joint analysis to try to reach consensus on policy formulation through a mutually agreed interpretation of the information about the proposed policy and its anticipated impacts.
■ Shared decision making involves direct participation and partial control over policy decisions, including policy formulation, implementation, monitoring, and evaluation. It is the highest level of participation in policy processes. It goes beyond giving vulnerable groups voice to giving them some measure of control.

Tools for Giving the Vulnerable Voice

A variety of tools and mechanisms is available for incorporating the voice of vulnerable groups into the policy process (table 5.1 and figure 5.3). Only a few of the most common are discussed here (for an introduction to a broader set of tools, see World Bank 2003c; Salmen and Kane 2006). Each policy process is set in a political, social, and cultural context that shapes which tools are most appropriate.

The first three tools—stakeholder analysis, institutional analysis, and social impact analysis—are often used together early in the policy process as part of policy analysis or diagnosis. They give vulnerable groups voice by seeking their views about potential policy impacts and by making them a key target group for policy diagnostics.

Stakeholder Analysis

Stakeholder analysis is used to determine the interests and influence of different groups in relation to a policy (for more information on stakeholder analysis, see Bianchi and Kossoudji 2001). It should precede policy design and be deepened as policy elements are elaborated.

Stakeholder analysis gives vulnerable groups voice by the following:
■ Identifying how they are a relevant stakeholder group whose inputs should be solicited and analyzed
■ Seeking their input in analyzing how their views may differ from other stakeholders and why
■ Analyzing the likelihood of their participation in coalitions to support change
■ Developing strategies for overcoming their opposition, such as giving them incentives in order to win their support for policies.

It was used in several World Bank–supported mine closure projects to create a system of checks, balances, and independent assessments to ensure that all actors followed the rules (Haney and others 2003).

Institutional Analysis

Institutional analysis unpacks the "black box" of decision making and implementation processes to identify how vulnerable groups can be incorporated into policy

TABLE 5.1
Tools for Participation in Policy Processes

Tool	What Is It?	What Can It Be Used For?	What Does It Reveal?	Key Elements
Stakeholder analysis	Systematic methodology that uses qualitative data to determine the interests and influence of different groups in relation to a reform or policy	Can be carried out for any type of development process; particularly amenable to sectoral reform, including policies. Basic stakeholder analysis should precede reform and policy design and should be consistently deepened as reform and policy elements are elaborated.	Assesses extent to which policy reform may spur political or social action; level of ownership among different groups; and perceptions of reform among different groups. Stakeholder analysis can be expanded into fuller political economy analysis that identifies affected groups and looks at their position with respect to policy; their influence on government; the likelihood of their participation in coalitions to support change; and strategies for overcoming opposition, such as compensating losers or delaying implementation.	• Background information on constraints to effective government policy making • Interviews with key informants that identify stakeholders relevant to sustainability of policy reform (participants should be drawn from a diverse groups of interests in order to limit bias) • Verification of assumptions about stakeholder influence and interest through survey work and quantitative analysis of secondary data
Institutional analysis	Analytical approach that uses qualitative methods to unpack the "black box" of decision making and implementation processes	Useful for poverty and social impact analysis regardless of reform type; particularly important for policy changes involving institutional reforms, such as decentralization of public services, utility reforms, land reforms, and social safety net reforms. Useful for policy design and implementation.	Analyzes institutions involved in design and implementation of reforms and identifies dynamic processes and potential constraints. Identifies characteristics and dynamic relationships between government agencies, NGOs, and firms that implement policy reform. One output is understanding of formal rules of the game and informal rules that govern actual behavior in decision-making processes (through process mapping of crucial resource flows, money, information).	• Background information on key stakeholders and organizational structures of relevant agencies • In-depth interviews or focus groups with key informants from government agencies, NGOs, and firms • Cross-referencing with other information to validate information from other sources

	Description	Application	What it reveals	Data sources/tools
Social impact analysis	Analytical framework to identify the range of social impacts and responses to policies by people and institutions, including those that are vulnerable or poor. Often undertaken in iterative manner; includes detailed information on social context for policy reform.	Can be used for many types of policy reforms. Has been used extensively for mining sector restructuring, parastatal privatization, and agricultural reforms with significant social impacts.	Reveals social and political context for policy reform: who is affected by the reform at what point in time; preferences and priorities of those affected by the policy; constraints to implementation of policy; and how people and institutions are likely to respond to policy reform, including validity of assumptions about how they will react or be affected by the policy. Also provides insight into coping mechanisms, social risks, and stakeholder views on most-appropriate means of mitigating negative impact of policies and potential effectiveness in local context.	• Open-ended community discussions • Interviews with key informants • Focus groups • Quantitative surveys • Observation • Ethnographic field research • Participatory rural appraisal Typically uses purposive surveys to collect quantitative information from sample representative of particular region or population groups relevant to a particular policy reform. Particularly useful when national household data do not exist or do not contain specific information needed to assess policy impacts.
Beneficiary assessment	Participatory assessment method and monitoring tool that incorporates direct consultation of those affected by and influencing policy reform. Similar to a participatory poverty assessment, it relies primarily on qualitative research, though with less emphasis on the use of visual techniques and community follow-up to the research process.	Has traditionally been used to evaluate projects or sectoral policy reforms in the health, education, infrastructure, social protection, and agricultural sectors; can be adapted to assess or monitor impact of some discrete policy interventions where transmission channels and affected groups are clearly defined. Used both to evaluate proposed reforms (to signal constraints to participation faced by target group) and to gain beneficiary feedback for ongoing policy reforms.	Reveals beneficiaries' perception of proposed policy and any mitigatory measures being considered. Provides insights into the likely reception reform will receive, as well as issues that may arise during implementation. Reaches the community level but is not focused exclusively on the poor or the community.	• Interviews • Focus group discussions, which in some cases have been combined with participatory rural appraisal tools • Direct and participant observation Although information collected is qualitative, it is analyzed quantitatively.

(continued)

TABLE 5.1
Tools for Participation in Policy Processes (continued)

Tool	What Is It?	What Can It Be Used For?	What Does It Reveal?	Key Elements
Participatory poverty assessment	Instrument for including the poor directly in discussions and debates on policies and priorities. Relies primarily on quantitative and qualitative, visual, and participatory rural appraisal techniques. Uses same data-collection techniques as beneficiary assessment, with sharper focus on consultation of the poor and broader set of policy issues affecting the poor.	Can be adapted to the analysis or monitoring of many policy reforms. Has been used extensively in public expenditure reforms that require priority setting and better understanding of reasons for lack of accountability or low service use for institutionally complex reforms (such as land reform, liberalization of markets, labor market reforms) and for better targeting safety nets. Could also be used to monitor local impact of macroeconomic policies.	Provides in-depth analysis of the views of the poor and their political, social, and institutional context; policy priorities of the poor; multidimensional dynamics of poverty and coping mechanisms; and constraints that could be overcome through public action to increase access to reform benefits, with focus on constraints for the poor.	Variety of flexible participatory methods that combine visual methods (mapping, matrices, diagrams) and verbal techniques (open-ended interviews, discussion groups) and emphasize exercises that facilitate information sharing, analysis, and action, with a goal of giving communities more control over the research process. May create opportunities or expectations of follow-up at the community level, such as development of community action plans, often supported by local government or NGOs.
Citizen report cards	Participatory survey that solicits client feedback on performance of public services. Combines qualitative and quantitative methods to collect useful demand-side data that can help improve performance of public services. Extensive media coverage and civil society advocacy allows tool to be used for public accountability.	Used in situations in which demand-side data, such as user perceptions on quality and satisfaction with public services, are absent. Provide avenue for citizens to signal public agencies and politicians on key reform areas and to spur competition among state-owned monopolies.	Provides feedback from users of services regarding issues such as availability of services; satisfaction with services; reliability and quality of services and indicators to measure them; responsiveness of service providers; hidden costs (corruption and support systems); willingness to pay; and quality of life.	• User-determined assessment criteria • Quantitative feedback on service delivery quality • Media involvement and broad public debate on process and survey results

Community scorecards	Community-based qualitative monitoring tool that draws on techniques of social audit, community monitoring, and citizen report cards. Increases empowerment and accountability through interface meeting between service providers and community that allows for immediate feedback.	Allow communities to monitor and evaluate performance of services, projects, and even government administrative units (such as district assemblies). Process allows for tracking of inputs or expenditures; monitoring of quality of services and projects; generation of benchmark performance criteria that can be used in resource allocation and budget decisions; comparison of performance across facilities and districts; generation of a direct feedback mechanism between providers and users; building of local capacity; and strengthening of citizen voice and community empowerment.	Provides information on how inputs or expenditures match entitlements and allocations at the local/facility level; criteria used by community and providers to assess performance; how both the community and providers score themselves on these criteria; anecdotal evidence on which these scores are based; and how assessments by the community and providers can be used to develop action plan for improvement.	• Community-designed and executed qualitative service assessment • Professionally facilitated public discussion of results

Source: Adapted from World Bank 2003c.

FIGURE 5.3
How Tools Fit into the Participatory Policy Cycle

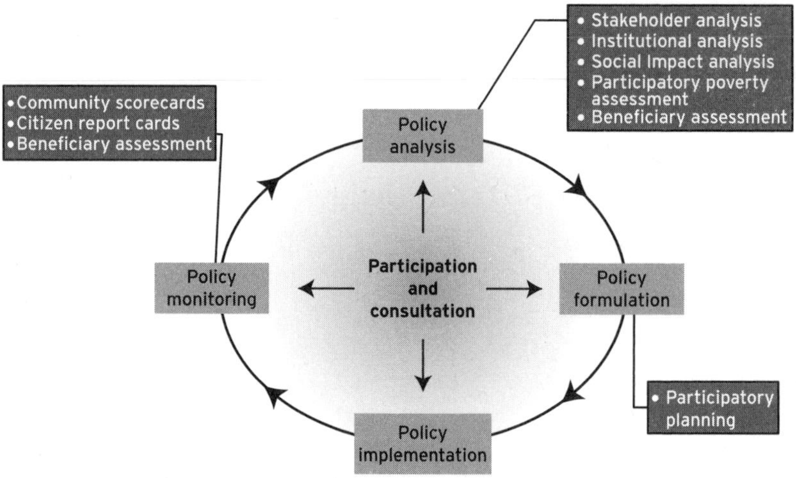

Source: Adapted from Shah and Youssef 2002.

processes (for more information on institutional analysis, see Brinkerhoff and
Crosby 2002). It is useful for policy design and implementation; it is particularly
important for policy changes involving institutional reforms, such as decentralization of public services. Institutional analysis does not directly give vulnerable
groups voice, but it is critical in identifying how they can play a role in decision-
making processes. It was used in the design of the World Bank–supported forestry
project in India to develop a more-inclusive legal and institutional framework
that allows forest-dependent communities to gain management control over their
natural resources (Kvam and Nordang 2004).

Social Impact Analysis

Social impact analysis identifies the range of social impacts and responses to
policies by people and institutions, including those that are vulnerable or poor
(for more information on social impact analysis, see Finsterbusch, Ingersoll, and
Llewellyn 1990; Becker 1997; Goldman 2000). Social impact analysis helps
ascertain the social and political context for policy reform; who is affected by the
policy; the preferences and priorities of those affected by the policy; constraints
to implementation of the policy; and how people and institutions are likely to
respond to policy reform, including whether assumptions on how they will react
or be affected by the policy were correct. It also provides insight into coping mechanisms and social risks, by seeking suggestions from stakeholders on the most
appropriate means to mitigate the negative impact of policies. It has been used

extensively for mining sector restructuring, parastatal privatization, and utility and agricultural policy reforms that have significant social impacts (Junge and others 2004; Beddies and De Soto 2005; Keener and Banerjee 2006).

Beneficiary Assessment and Participatory Poverty Assessment

The next two tools can be used as part of policy analysis or policy monitoring. As part of policy analysis, they are often used in conjunction with stakeholder analysis or social impact analysis. They give vulnerable groups voice by making them a key target group for policy diagnostics and by seeking their feedback on actual impacts of policy implementation.

Beneficiary assessment is a participatory assessment method and monitoring tool that incorporates direct consultation of those affected by and influencing a policy (for more information on beneficiary assessment, see Salmen and Amelga 1998; Salmen 2002). Beneficiary assessment of vulnerable groups could give them voice by identifying their constraints to participation and obtaining their feedback on the policy, the problem the policy addresses, and any mitigatory measures being considered. Five beneficiary assessments in the Zambia Social Recovery Project studied the socioeconomic profile of project beneficiaries and assessed coping strategies during structural adjustment and drought. Subsequent beneficiary assessments assessed the impact of changes introduced by social fund management in response to recommendations from earlier beneficiary assessments (Jones and Owen 1998).

A participatory poverty assessment is an instrument for including the poor directly in discussions of policies and priorities (for more information on participatory poverty assessment, see Norton and others 2001; Robb 2002). Participatory poverty assessments are designed in consultation with policy makers and civil society groups. Unlike a household survey, which consists of a predetermined set of questions, a participatory poverty assessment uses a variety of flexible participatory methods. These methods combine visual techniques (mapping, matrices, diagrams) and verbal techniques (open-ended interviews, discussion groups); they emphasize exercises that facilitate information sharing, analysis, and action.

Participatory poverty assessments go beyond the household unit of traditional surveys to focus on individuals, intrahousehold dynamics, social groups, and community relationships. They give vulnerable groups voice by analyzing their views and their political, social, and institutional context, as well as their policy priorities and coping mechanisms; identifying constraints that could be overcome through public action to increase their access to policy benefits; and increasing their capacity to analyze and monitor poverty. The Vietnam participatory poverty assessment prioritized interventions that increased and stabilized agricultural and off-farm incomes; increased access of poor migrants to government services; improved the targeting of interventions toward poor households; minimized preferential access to services and resources by better-connected households; and improved the transparency of beneficiary identification for government programs (Turk 2001; Robb 2003).

Citizen Report Cards and Community Scorecards

The last two tools are normally used as part of policy monitoring. They give vulnerable groups voice by seeking their feedback on the impacts of policy implementation. These tools are useful for giving voice to citizens in a way that can influence actions by both the public and private sectors.

A citizen report card is a participatory survey that solicits client feedback on the performance of public services (for examples and more information, see World Bank 2001; Paul 2002; Public Affairs Centre 2002; Wagle, Singh, and Shah 2004). They are used in situations in which demand-side data, such as user perceptions on quality and satisfaction with public services, are absent. By collecting and aggregating user feedback, they provide an avenue for citizens to signal public agencies and politicians on key policy areas. They also create competition among state-owned monopolies. They give vulnerable groups voice to the extent that they are included among those who are asked to provide user feedback on government or service provider performance. While typically used to monitor public service performance, citizen report cards can also be used to monitor policy impacts.

A community scorecard is a community-based qualitative monitoring tool that draws on techniques of social audit, community monitoring, and citizen report cards.[6] The process allows for monitoring of the quality of services and projects; generation of benchmark performance criteria that can be used in resource allocation and budget decisions; comparison of performance across facilities and districts; creation of a direct feedback mechanism between providers and users; local capacity building; and strengthening of citizen voice and community empowerment. Although community scorecards are usually used for project monitoring, they can be adapted to monitor policies. Citizen report cards and community scorecards are part of a broader approach to promoting good governance and making governments accountable and responsive to its citizens known as *social accountability* (for an introduction to the theory and practice of social accountability, see Malena, with Forster and Singh 2004).

Case Studies

This section examines two case studies in which vulnerable groups were given voice in shaping environmental policy. Each example describes the political economy context of the policy process, how vulnerable groups participated in the policy process (including what tools were used), what the findings were of that process, how their voice was ensured, what impacts their voice had on policy, and what impacts the policy had on vulnerable groups.

Participatory Budgeting and Environmental Management in Porto Alegre, Brazil

Porto Alegre is one of the best-known cases in which vulnerable groups have been given voice in municipal-level planning. Porto Alegre is the capital of Brazil's southernmost state, Rio Grande do Sul. It has the highest standard of living and the longest life expectancy of any Brazilian metropolitan center (Menegat 2002). Virtually all its people have water piped to their homes, and most have good-quality sanitation and drainage. The garbage collection system reaches virtually all households and has included separate collection of recyclables since 1990. Other programs enforce industrial pollution control, reduce motor vehicle emissions, and ensure the reuse of organic wastes from parks and restaurants.

Southern Brazil has historically demonstrated a strong democratic political culture and resistance to authoritarianism. When direct elections for capital city mayors were reestablished, Porto Alegre was one of the few municipalities the new Workers Party won in the 1988 elections. The Workers Party had a very clear agenda, which advocated direct participation, redistribution, and a reduction in corruption.

The 1989 municipal elections brought to power a progressive mayor who realized the potential for opening up to citizen's scrutiny the dire fiscal situation inherited from the previous administration and engaging citizens in the prioritization of local needs with respect to scarce municipal resources (Wagle and Shah 2003). Facing extremely high expectations from constituents, financial meltdown, and threats from municipal councilors, the mayor shifted the local governance culture from the traditional executive-legislative focus toward the people. This created significant political tensions, not only among the opposition but also among segments of the left, which were not keen to change the political culture and power relations.

Process. Participatory budgeting completely reversed the traditional patronage approach that characterizes public administration in most Brazilian cities. The process of discussion and decision making follows an annual cycle of two main stages. The first involves defining priorities and proposals for public spending in plenary assemblies, in which all citizens can participate. The second involves drawing up the budget proposal and expenditure plan. The priorities and proposals approved by citizens are supposed to be sufficiently developed for submission to the state legislature as the municipal budget. In 2000 the participatory budgeting process involved about 30,000 citizens, ensuring that public interventions corresponded with the priorities of the population (Menegat 2002). Since 1989, almost 15 percent of the population has participated in at least one budget event.

The participatory budgeting cycle usually starts with the municipality organizing information meetings. Citizens are provided with necessary information, such as budget rules and procedures, implementation status of the current budget, and government expenditure priorities. The first stage consists of two large rounds of general and sectoral plenary assemblies. Citizens can participate in all events, during which they have the opportunity to present their requests and proposals for the annual municipal budget for their district or for a specific sector. Between the two rounds is an interim phase, which consists of numerous more-specific meetings in each of the municipality's 16 districts and on five sectoral themes. These meetings are coordinated and facilitated by the delegates elected to the district and sectoral assemblies. They allow communities to discuss their needs and priorities in greater depth.

The second round of meetings is coordinated by the participatory budgeting council. During this round, citizens' priorities are identified and voted on. Citizen representatives undertake capacity-building activities to improve citizens' understanding of public budgeting, fiscal mechanisms, and service and investment planning, as well as to enhance their consensus-building and conflict resolution skills. The representatives usually conduct field inspections to validate the priorities selected during the regional meetings. The government conducts technical and financial feasibility studies for each priority presented at the budget council. After intensive debates between the government and citizens' delegates, the final proposal is voted on by the budget council and presented to the mayor for budget consolidation and submission to the municipal legislative council.

Grassroots groups mobilize vulnerable people so that they can have a significant presence in the participatory budgeting process and elect their own representative on the participatory budgeting council, the ultimate level of decision making. These groups implement an outreach campaign, help strengthen leadership among vulnerable people, and advocate their needs in policy and budgeting decision making. Because the process is time intensive, the most vulnerable groups often participate less than other social actors.

Findings and impact. Once citizens had become involved in decision making for the municipal budget, it was recognized that urban planning and management, in particular environmental management, needed to be fundamentally revised. It was clear that a purely physical planning model that operated in isolation from the key actors was incompatible with participatory democracy. The planning authority needed to adopt democratic decision-making processes in its relations with other government bodies, with other institutions, and with citizens. The need was identified for a system of knowledge and information of the urban and natural environments that was accessible to all stakeholders, including planners, politicians, institutions, and citizens. The production of this knowledge itself required fundamental changes in the way information was collected, compiled, and validated.

The system for integrated environmental management is based on four interrelated components: citizen participation, public environmental management programs, comprehensive knowledge of Porto Alegre's natural and built environments, and environmental education (Menegat 2002). Citizens participate in environmental policy making through either the municipal councils or the assemblies within participatory budgeting.

Several environmental programs have been undertaken to implement the more democratic and participatory environmental policy (Menegat 2002). Urban growth has caused the degradation of public spaces, particularly green areas. Two ongoing programs aim to address the problem by involving citizens in their management. The first is the Green Area Adoption Scheme, in which a partner institution "adopts" a square or park and commits to undertake the maintenance and gardening. The second is the City Square Councils Program, which brings together interested citizens, civil society organizations, and businesses, who organize the maintenance and gardening of a particular city square and define the rules for its use. Tree planting along streets is one of the specific measures for environmental protection set out in municipal legislation. Special programs were started in 1990 for pollution control of oil, water, and the atmosphere. The growing problem of refuse disposal has been mitigated through the integrated solid waste management program, which has both reduced the quantity of waste disposed of in landfills (thus increasing their life) and generated income from recycling.

For citizens to participate in urban and environmental management in a meaningful way, they need information and knowledge about the natural and built environments. In Porto Alegre such information was disseminated principally through the *Environmental Atlas of Porto Alegre* (Menegat and others 1998). The *Atlas* uses plain language to present the relationships between the local, regional, and global environments, from the geosphere, hydrosphere, biosphere, and atmosphere to the technosphere and urbansphere. It provides a solid foundation of scientific knowledge that can be used to produce scenarios and forecasts on which to base the city's environmental monitoring, legislation, and licensing. The establishment of environmental directives for various scenarios also means that specific issues raised by citizens can be responded to more quickly, as the information will be more readily available. Following the publication of the *Environmental Atlas of Porto Alegre*, an environmental education program was established.

Participatory budgeting had instrumental value in promoting environmental education, programs, and goals. It also had intrinsic value in bringing about a fundamental change in the political culture of Porto Alegre. This change signified an end to the traditional top-down style of decision making, the redefinition of public priorities in line with citizens' views, and the transition to an inclusive city. The best illustration of the political impact of participatory budgeting in Porto Alegre is that after 16 years in power, the Workers Party lost the 2004 election to a coalition composed of all the opposition parties that were originally against but eventually supported participatory budgeting. The new mayor won election because he acknowledged the importance of participatory budgeting and the need to improve local governance mechanisms even further.

A rigorous impact assessment carried out by the World Bank (Baiocchi and others 2006) compares various indicators in municipalities that used participatory budgeting and municipalities that did not. It finds that participatory budgeting had an impact, particularly for the poorest citizens, on the level of poverty and that it increased the voice of vulnerable groups. Marquetti (2000) shows that participatory budgeting in Porto Alegre increased the resources allocated to services that benefit poor and vulnerable groups.

Why the voice of vulnerable groups was heard. The voice of vulnerable groups was heard in Porto Alegre because a grassroots movement won power through elections and the elected government was accountable and responsive to the poor and vulnerable. The social movement in Brazil demanded direct participation in decision making and control over resources. Once political freedom was reestablished in Brazil, many activists who were in the social movement joined political parties, mainly the Workers Party, bringing the social movement into the political arena. Participatory budgeting itself has become a powerful force in shaping the politics of Porto Alegre, so much so that the Workers Party lost power in 2004, at least partly because it had not gone far enough in the view of vulnerable groups.

Policy toward Pastoralists in Mongolia

Under the communist regime that ruled the country until 1990, Mongolia made great progress in improving human development indicators and virtually eliminating poverty (Mearns 2004). Innovative service delivery mechanisms to nomadic pastoralists achieved almost universal coverage of primary health care and basic education services. Life expectancy increased from 47 in 1960 to 63 in 1990. Adult literacy reached 97 percent.

The sudden loss of Soviet subsidies led to a one-third decline in gross domestic product (GDP). The 1990s saw political and economic transition and a rapid rise in asset and income inequality. The economic transition led to a dramatic shakeout of labor from uneconomic state-owned enterprises that was absorbed largely by the extensive livestock sector in rural areas.

By the late 1990s, herders accounted for more than a third of the population and half the active labor force. Pressures on common pastures grew, increasing violence, livestock theft, and conflict over pastures. In 1998–99 export earnings fell because of the collapse in prices of Mongolia's three main exports: gold, copper, and cashmere. More than a third of the population was defined as living under the poverty line.

Three successive years (1999–2002) of drought and harsh winters led to high livestock mortality and prompted two sorts of responses. The first was to recognize that household-level vulnerability to such risk episodes had been exacerbated by neglect at the level of public policy and investment in the livestock sector and other support frameworks over the preceding decade. The second was to reinvigorate approaches to rural development, with particularly attention to pastoral risk management, including ways of facilitating livestock mobility, both seasonally and in response to episodic risk.

Process. The Mongolia Participatory Poverty Assessment (the Participatory Living Standards Assessment [PLSA]), was conducted against the backdrop of the general election campaign in 2000. The incumbent democratic coalition government felt vulnerable to criticisms that poverty had not been declining quickly enough after economic growth picked up in the mid-late 1990s. The former communist party, the Mongolian People's Revolutionary Party (MPRP), ran on a platform that emphasized, among other things, the importance of a poverty-reduction-with-growth agenda. Poverty was therefore a highly charged political issue in 2000, as well as the foremost developmental challenge facing the country. The PLSA (National Statistical Office of Mongolia and World Bank 2001) was explicitly designed as an attempt to elicit the voices of the poor and vulnerable and to create space for their voices to be heard in framing a new national poverty reduction program. The PLSA set out to analyze not only poverty but also its wider context.

The PLSA focused on social differentiation, vulnerability, and access to assets as key issues for more-secure and sustainable livelihoods (Dulamdary and others 2001). It was the first exercise of its kind in Mongolia to use participatory learning and action methods to broader and deepen understanding of poverty at the national level.

The assessment was based on the sustainable livelihoods approach, which emphasizes the range of capital assets (natural, human, social, physical, and financial) that people draw upon in pursuing diverse livelihood strategies. It used a range of participatory learning and action methods, including matrix ranking and scoring, wealth and well-being ranking, semi-structured interviews, and focus groups.

The PLSA began with participants' own understandings of well-being. This led to an analysis and discussion of differences among households within communities. More than 2,000 people participated in the PLSA as informants or focus group members. Presentations of preliminary results were made at national conferences, workshops, and briefings to members of parliament, senior government officials, NGOs, civil society groups, and donor agencies. A competition among various media was held to promote public debate on poverty and public and private actions.

Findings and impact. The PLSA highlighted the multiple sources of insecurity and vulnerability that had emerged as a result of the privatization of state-owned enterprises and pastoral collectives. It also outlined the coping strategies of poor communities. Priorities for action included the following:

- Recognize a view of ill-being that is broader than poverty and includes alcohol abuse, crime, and domestic violence.
- Shift from creating employment to promoting people's capacity to secure their own livelihoods.
- Reduce vulnerability to risks through social networks, life skills, and innovative microfinance products, including livestock insurance.
- Reduce risk to pastoral livestock production through restoration of pastoral mobility, community-based pasture land management, and livelihood diversification.
- Improve the quality and effectiveness of social services and infrastructure as a basis for thriving local economies.
- Increase public access to information, and give citizens greater voice and influence over public spending.

The PLSA led to a broader public discourse on poverty and to increased understanding of the multiple dimensions, causes, and consequences of impoverishment and vulnerability. Mongolia's Poverty Reduction Strategy Paper and rural development strategy reflected many of the findings of the PLSA and formed the basis for the government's Household Livelihoods Capacity Support Program (Government of Mongolia 2003).

The PLSA also helped frame the design of the new national poverty program in terms of a sustainable livelihoods approach. The key components include pastoral risk management, sustainable microfinance services to underserved groups such as herders, and a more community-driven approach to identifying and managing investments in basic infrastructure, all of which give voice to vulnerable groups.

Perhaps the clearest example of how the voices of vulnerable groups influenced policy is the land law and policy toward pasture land tenure. Participation in the PLSA created the opportunity to enshrine in law the rights of poor and vulnerable herders. The revised Land Law prohibited privatization of pasture land and endorsed a more socially inclusive, common-pool approach to pasture land management. Both are of central importance for the livelihoods of herders, whom of whom are poor and vulnerable.

Why the voice of vulnerable groups was heard. The voice of vulnerable groups may have been heard and taken into account in the PLSA because it served the political interests of the MPRP to demonstrate its commitment to following through on its campaign promises of pro-poor action.[7] A sufficient coalition of interests—in government, parliament, NGOs, think tanks, and other civil society institutions—was forged that heard the voice of vulnerable groups to permit some enabling progressive policies (such as some aspects of the land law and the Household Livelihoods Capacity Support Program) to find a foothold. Some highly committed public figures in these institutions have a genuine interest in progressive social change and believe in the power of public action to help bring it about. These interests were able to stand up to opposition from very powerful individuals (such as the prime minister).

Conclusion

Giving vulnerable groups a voice in policy formulation yields many benefits. Policies are better informed, understood, and supported. The vulnerable are empowered and able to influence policies. They also benefit from helping make government more inclusive and accountable to their needs. Giving voice to vulnerable groups is often a virtuous circle, in which empowerment leads to better policies and policy impact, which lead to further empowerment and other benefits for vulnerable groups. The key is how to get the virtuous circle started.

The vulnerable are not simply poor. They also suffer from insecurity, isolation, lack of dignity, lack of access to information, and powerlessness. They are dependent on natural resources for their livelihoods, live in unhealthy environments, are more vulnerable to waste management and toxins, and are more prone to the impact of natural disasters and climate change. If the enabling environment is weak, participation may remain a one-way information flow. Social accountability and transparency require participation that goes beyond one-way information sharing to joint-decision making. In this way citizens are able to understand, engage with, and influence policy making and implementation.

Adaptive management and inclusive decision making create opportunities for giving the vulnerable voice. An enabling environment—including the freedom to associate, the resources to facilitate voice, the ability to exercise voice, and the existence of spaces for voice—are critical. As both case study examples show, without political support from those in power, the voice of vulnerable groups would not have been heard, much less acted upon.

Within the policy process, voice can be incorporated in policy analysis, formulation, implementation, and monitoring. Participatory planning, budgeting, and decision making have been effective means for making the voice of vulnerable groups heard. Some of the most common tools for giving voice include stakeholder analysis, institutional analysis, social impact analysis, participatory poverty assessment, beneficiary assessment, participatory planning, citizen report cards, and community scorecards. These tools are techniques: what is important is the willingness and capacity to give vulnerable groups voice, which usually requires political power or support from those in power to complement social mobilization from below.

Giving vulnerable groups voice has had a significant impact. It has led to a broader public discourse on poverty and to increased understanding of the multiple dimensions, causes, and consequences of impoverishment and vulnerability.

It has helped shape poverty reduction and sectoral strategies (including environmental strategies, policies, and programs) and led policy makers to examine the impacts of policies and reforms on the poor. It has prioritized interventions, improved access to and the quality of government services, improved the targeting of interventions toward vulnerable groups, and enhanced the capacity to carry out participatory research on policy issues. Direct consultation with vulnerable groups can present opportunities for a more-open dialogue and greater understanding between those in power and vulnerable groups.

Participation of vulnerable groups in policy processes is still in its infancy: participation in policy analysis, formulation, and monitoring has not been matched by participation in policy implementation. Most participation has been in the social sectors (health, education, social protection) or governance processes (budgets, decentralization). Only more recently has it expanded to sectors in which environmental considerations are more important (energy, agriculture, natural resource management). While a good start has been made, much remains to be done to improve the consistency, quality, and effectiveness of the participation of vulnerable groups.

Notes

1 Robb and Scott (2001), Tikare and others (2001), and Shah and Youssef (2002) summarize these arguments.

2 For information on evidence-based policy making using participatory approaches to include vulnerable groups and civil society more broadly, see Pollard (2005).

3 It is necessary to have an appreciation of the impact of reforms on groups other than the poor and vulnerable, in order to understand why certain groups may want to influence or block certain policy decisions.

4 Social accountability refers to the broad range of actions and mechanisms, beyond voting, that citizens can use to hold the state to account, as well as actions on the part of government, civil society, media, and other societal actors that promote or facilitate these efforts.

5 A policy can be considered fair if individuals would support it if they did not know in advance what their position in society would be (Rawls 1971).

6 For more information on a community scorecard process, see World Bank (2004a). For an example, see Dedu and Kajubi (2005).

7 For more information, see World Bank (2007).

References

Ackerman, J. 2005. *Social Accountability in the Public Sector: A Conceptual Discussion.* Social Development Paper 82, World Bank, Washington, DC.

Baiocchi, G., P. Heller, S. Chaudhuri, and M. K. Silva. 2006. "Evaluating Empowerment: Participatory Budgeting in Brazilian Municipalities." In *Empowerment in Practice: From Analysis to Implementation*, eds. R. Alsop, M. Bertelsen, and J. Holland. Washington, DC: World Bank.

Becker, H. 1997. *Social Impact Assessment*. London: University College Press.

Beddies, S., and H. De Soto. 2005. *Poverty and Social Impact Analysis (PSIA) of the Decentralization and Water Sector Privatization in Albania*. Washington, DC: World Bank.

Bianchi, R., and S. Kossoudji. 2001. *Interest Groups and Organizations as Stakeholders*. Social Development Paper 35, World Bank, Washington, DC.

Blackburn, J., R. Chambers, and J. Gaventa. 2000. "Mainstreaming Participation in Development." Working Paper, World Bank, Operations Evaluation Department, Washington, DC.

Brinkerhoff, D., and B. L. Crosby. 2002. *Managing Policy Reform: Concepts and Tools for Decision-Makers in Developing and Transition Countries*. Bloomfield, CT: Kumarian Press.

Chambers, R. 1983. *Putting the First Last*. London: Longman.

———. 2007. "Who Counts? The Quiet Revolution of Participation and Numbers." IDS Working Paper 296, Institute of Development Studies, Sussex, United Kingdom.

Dedu, G., and G. Kajubi. 2005. "The Community Scorecard Process in The Gambia." Social Development Note 100, World Bank, Washington, DC.

DeWind, J., and D. H. Kinley. 1988. *Aiding Migration: The Impact of International Development Assistance on Haiti*. Boulder, CO: Westview Press.

Dulamdary, E., M. Shah, and R. Mearns, with B. Enkhbat and L. Ganzaya. 2001. "Mongolia: Participatory Living Standards Assessment," Report prepared for the Donors' Consultative Group Meeting, Paris, May 15–16. National Statistics Office, Ulaanbaatar, Mongolia, and World Bank, Washington, DC.

Feldman, M., and A. Khademian. 2000. "Management for Inclusion: Balancing Control with Participation." *International Public Management Journal* 3 (2): 149–68.

———. 2005. "Inclusive Management: Building Relationships with the Public." Paper 04-12, Center for the Study of Democracy, University of California–Irvine.

Finsterbusch, K., J. Ingersoll, and L. Llewellyn. 1990. *Methods for Social Analysis in Developing Countries*. San Francisco: Westview Press.

Goldman, L. R., ed. 2000. *Social Impact Analysis: An Applied Anthropology Manual*. Oxford: Berg Press.

Government of Mongolia. 2003. *Economic Growth Support and Poverty Reduction Strategy*. Prime Minister's Office, Ulaanbaatar.

Grindle, M. S., and J. W. Thomas. 1991. *Public Choices and Policy Change: The Political Economy of Reform in Developing Countries*. Baltimore, MD: Johns Hopkins University Press.

Haney, M., M. Shkaratan, V. Kabalina, V. Paniotto, and C. Rughinis. 2003. *Mine Closure and Its Impact on the Community: Five Years after Mine Closure in Romania, Russia, and Ukraine*. Social Development Paper 42, World Bank, Washington, DC.

Holland, J., and J. Blackburn. 1998. *Whose Voice? Participatory Research and Policy Change*. Intermediate Technology Development Group, London.

IEA (International Energy Agency). 2006. *International Energy Outlook 2006*. Paris: Organisation for Economic Co-operation and Development/IEA.

Jones, B., and D. Owen.1998. "Beneficiary Assessment for Monitoring: The Zambia Social Recovery Project." Social Development Note 36, World Bank, Washington, DC.

Junge, N., T. Pushak, J. Lampietti, N. Dudwick, and K. Van den Berg. 2004. *Sharing Power: Lessons Learned from the Reform and Privatization of Moldova's Electricity Sector*. World Bank, Washington, DC.

Kanji, N., and S. Ware. 2002. "Trade Liberalization, Poverty and Livelihoods: Understanding the Linkages." Review prepared for the African Policy and Economic Department, Department for International Development, London.

Keener, S., and S. Banerjee. 2006. "Ghana: Electricity Tariff Reform." In *Poverty and Social Impact Analysis of Reforms*, eds. A. Coudouel, A. Dani, and S. Paternostro. Washington, DC: World Bank.

Kvam, R, and H. Nordang. 2004. *The Jharkand Participatory Forest Management Project: Social Assessment for Inclusion, Cohesion, and Accountability.* World Bank, Washington, DC.

Leitner, K. 2005. "Heath and Environment: A View from WHO." In *Environment Matters.* Washington, DC: World Bank.

Malena, C., with R. Forster and J. Singh. 2004. "Social Accountability: An Introduction to the Concept and Emerging Practice." Social Development Paper 74, World Bank, Washington, DC.

Marquetti, A. 2000. "Participatory Budgeting in Porto Alegre." *Indicator S A*, Johannesburg, 17 (4): 71–78.

Mearns, R. 2004. "Sustaining Livelihoods on Mongolia's Pastoral Commons: Insights from a Participatory Poverty Assessment." *Development and Change* 35 (1): 107–39.

Menegat, R. 2002. "Participatory Democracy and Sustainable Development: Integrated Urban Environmental Management in Porto Alegre, Brazil." *Environment and Urbanization* 14 (2): 181–206.

Menegat, Rualdo, Maria Luiza Porto, Clovis Carlos Carrazo, and Luís Alberto Dávila Fernandes.1998. *Environmental Atlas of Porto Alegre.* Porto Alegre, Brazil: Edufrgs.

Narayan, D., R. Chambers, M. Shah, and P. Petesch. 2001. *Voices of the Poor: Crying Out for Change.* Washington, DC: World Bank.

National Statistical Office of Mongolia, and World Bank. 2001. *Mongolia Participatory Living Standards Assessment 2000.* Ulaanbataar, Mongolia.

Norton, A., B. Bird, K. Brock, M. Kakande, and C. Turk. 2001. *A Rough Guide to Participatory Poverty Assessments: An Introduction to Theory and Practice.* London: ODI Publications.

Ostro, B. 1994. "Estimating the Health Effects of Air Pollutants: A Methodology with Application to Jakarta." Policy Research Working Paper 1301, World Bank, Washington, DC.

Paul, S. 2002. *Holding the State to Account: Citizen Monitoring in Action.* Bangalore, India: Books for Change.

Pollard, A. 2005. *How Civil Society Organisations Use Evidence to Influence Policy Processes: An Annotated Bibliography.* London: Overseas Development Institute.

Public Affairs Centre. 2002. *The State of Karnataka's Public Services: Benchmarks for the New Millennium.* Bangalore, India.

Rawls, J. 1971. *A Theory of Justice.* Cambridge, MA: Belknap Press of Harvard University Press.

Robb, C. 2000. "How the Poor Can Have a Voice in Government Policy." *Finance and Development* 37 (4), International Monetary Fund, Washington, DC.

———. 2002. *Can the Poor Influence Policy? Participatory Poverty Assessments in the Developing World.* Washington, DC: International Monetary Fund and World Bank.

———. 2003. "Poverty and Social Impact Analysis: Linking Macroeconomic Policies to Poverty Outcomes: Summary of Early Experiences." IMF Working Paper WP/03/43, International Monetary Fund, Washington, DC.

Robb, C., and A. Scott. 2001. "Reviewing Some Early Poverty Reduction Strategy Papers in Africa." IMF Policy Discussion Paper PDP/01/5, International Monetary Fund, Washington, DC.

Salmen, L. 1995. "Listening to the People." *Finance and Development* 32 (2): 44–48: International Monetary Fund, Washington, DC.

———. 2002. *Beneficiary Assessment: An Approach Described.* Social Development Paper 10, World Bank, Washington, DC.

Salmen, L, and M. Amelga. 1998. *Implementing Beneficiary Assessment in Education: A Guide for Practitioners.* Social Development Paper 25, World Bank, Washington, DC.

Salmen, L., and E. Kane. 2006. *Bridging Diversity: Participatory Learning for Responsive Development.* Washington, DC: World Bank.

Shah, P., and D. Youssef. 2002. "Voices and Choices at a Macro Level: Participation in Country-Owned Poverty Reduction Strategies." Action Learning Program Dissemination Series 1. World Bank, Social Development Department, Washington, DC.

Subbarao, K., A. Bonnerjee, J. Braithwaite, S. Carvalho, K. Ezemenari, C. Graham, and A. Thompson. 1997. *Safety Net Programs and Poverty Reduction: Lessons from Cross-Country Experience.* Washington, DC: World Bank.

Thindwa, J., C. Monico, and W. Reuben. 2003. "Enabling Environments for Civic Engagement in PRSP Countries." Social Development Note 82, World Bank, Washington, DC.

Tikare, S., D. Youssef, P. Donnelly-Roark, and P. Shah. 2001. "Organizing Participatory Processes in the Poverty Reduction Strategy Process." In *PRSP Sourcebook.* Washington, DC: World Bank.

Turk, C. 2001. "Linking Participatory Poverty Assessments to Policy and Policy Making: Experience from Vietnam." Policy Research Working Paper 2526, World Bank, Washington, DC.

Wagle, S., and P. Shah. 2003. "Porto Alegre, Brazil: Participatory Approaches in Budgeting and Public Expenditure Management." Social Development Note 71, World Bank, Washington, DC.

Wagle, S., J. Singh, and P. Shah. 2004. "Citizen Report Card Surveys: A Note on the Concept and Methodology." Social Development Note 91, World Bank, Washington, DC.

Whittingham, E., J. Campbell, and Philip Townsley. 2003. *Poverty and Reefs.* Department for International Development, London.

World Bank. 2001. *Making Sustainable Commitments: An Environment Strategy for the World Bank.* Washington, DC: World Bank.

———. 2002. *World Development Report 2003: Sustainable Development in a Dynamic World.* Washington, DC: World Bank.

———. 2003a. *Cornerstones for Conservation: World Bank Assistance for Protected Areas.* Washington, DC: World Bank.

———. 2003b. *Indonesia Kerinci Seblat Integrated Conservation and Development Project: Implementation Completion Report.* Washington, DC: Indonesia Country Unit, World Bank.

———. 2003c. *A User's Guide to Poverty and Social Impact Analysis.* Washington, DC: World Bank, Poverty Reduction Group and Social Development Department.

———. 2004a. "Community Scorecard Process: A Short Note on the General Methodology for Implementation." Social Development Department, World Bank, Washington, DC.

———. 2004b. *Responsible Growth for the New Millennium.* Washington, DC: World Bank.

————. 2005. "Indonesia Coral Reef Rehabilitation and Management Project (COREMAP): Implementation Completion Report." World Bank, Indonesia Country Office.

————. 2007. "An Enabling Environment for Social Accountability in Mongolia." Report 32584 MN, World Bank, Social Development Department, Washington, DC.

World Bank, and DFID (Department for International Development). 2005. *Tools for Institutional, Political and Social Analysis (TIPS) in Poverty and Social Impact Analysis (PSIA): A Sourcebook for Commissioners and Practitioners*, vol. 1. Washington, DC: World Bank, and London: DFID.

World Health Organization. 2002. *World Health Report*. Geneva: WHO.

C H A P T E R   6

Building and Reinforcing Social Accountability for Improved Environmental Governance

Harry Blair

IF DEVELOPING COUNTRIES ARE TO CRAFT AND NURTURE
sustainable policy initiatives that can address externalities in ways that will help
the environment, they will need long-term constituencies that want to support
such policies and can hold policy makers accountable for their performance in
implementing them. Transparency will be the critical quality in the policy process
needed for these constituencies to demand accountability from policy makers.
This chapter explores the three key variables of accountability, transparency, and
long-term constituency building as crucial factors in dealing with externalities in
order to protect the environment in developing countries.

The chapter begins by defining some key concepts before constructing a theoretical framework that brings the essential variables together as part of the policy
process. The third section presents two case studies illustrating the three variables
in an environmental context. The last section identifies patterns and themes
observed in the case studies.

Harry Blair is a senior research scholar and lecturer in political science at Yale University. This
chapter benefited greatly from comments generously provided by Kulsum Ahmed, Sameer
Akbar, Giovanna Dore, and Jeff Thindwa, all of the World Bank.

Definitions and Causal Linkages

Several concepts are key to understanding how accountability is used to effect changes in environmental policy. Each of these concepts is defined here.

Accountability

Except for a few autocrats and top-level elites in corrupt systems, all political actors are accountable or answerable to someone—the question is, to whom? Only to a dictator? Just to elites? To the military? To voters? To the rule of law? In a very general way, the advance of democracy can be gauged by the number and kinds of quarters to which actors are accountable: the more developed the democracy, the greater and more widespread the accountability.[1]

In "electoral democracies," accountability comes only through elections (Diamond 1999); in full "liberal democracies," accountability mechanisms include legislative oversight, civil society advocacy, legal redress (rule of law), a free and active media, and shared power (between branches of government). In a democratic system, to be sure, elections form the ultimate defense against state mismanagement, offering the citizenry a chance to change system managers. In Tocqueville's famous formulation, elections are why "the great privilege of the Americans is to be able to make reparable mistakes," which, he believed, inevitably occur in a democratic setup (Tocqueville 2000: 216). But elections represent a very crude mechanism for accountability; they occur only at infrequent intervals and allow no possibility for the citizenry to exercise anything more than the crudest policy guidance. To hold the state to account for particular policies or actions (or inactions), more finely tuned instruments are needed that can function between elections to make specific demands and intervene in the policy process for specific purposes. This is where constituencies come in, for they are the actors making these demands. But without transparency, they will not know what to demand or who to demand it from.

Social accountability refers to the accountability of the state to the society as a whole (as opposed to some individual sector of society). This chapter focuses on social accountability for environmental protection.

Transparency

In this era of Transparency International as a major think tank in the development community, *transparency* tends to be thought as corruption's antonym—that is, honesty and probity in the public sector. Countries such as Bangladesh and Chad, which score at the bottom of Transparency International's annual Corruption Perceptions Index are seen as the essence of state corruption, while those at the top (Iceland and Finland) are perceived as being the most transparent.

In this chapter, *transparency* is used differently: to mean openness and accessibility of state decision-making processes to public scrutiny. The processes themselves may not always be completely honest, but if they are transparent, their degree of probity or venality will be open to public view. In every political system, dishonesty is bound to occur from time to time, but if the system is transparent and civil society organizations (CSOs) and especially the media are exercising a watchdog function, the malfeasance will be discovered and publicized. If other aspects explored in this chapter are working properly, more transparency should lead to more system probity.

Constituency

A true constituency is a group genuinely involved in public policy decision making, on both the input and output sides. Constituencies make claims on policy decision making; if successful, they benefit from policy implementation. They potentially include state actors (the military, the bureaucracy); voluntary advocacy groups (trade unions, environmentalists, professional associations, businesses, business chambers); ascriptively based groups (ethnic minorities, religious groups, women, linguistic communities); occupational categories (landlords, sharecroppers, forest loggers, bus operators); residents of a neighborhood or locality (slum inhabitants, forest dwellers, people living near a factory); and ultimately the citizenry at large, especially at times of systemic stress.[2]

Not every constituency contributes positively to the overall public weal; some can be damaging. Criminal gangs, timber thieves, environmental polluters, and fanatical religious groups are all constituencies, sometimes powerful ones.

Not every social grouping is a constituency. A constituency has achieved a consciousness of its identity and has developed an organization of some sort to advance its interests. In contrast, a social grouping exists only as a taxonomic category. To paraphrase a Marxist adage, a constituency arises when a group existing in itself has become a group organized for itself.[3]

The state as a set of constituencies. In this chapter, the "state" includes the public sector at all levels. Its three branches—executive, legislature, and judiciary—can also act as constituencies, making demands on the state (such as patronage funds for legislators feeling jealous of local government units). Two other state constituencies—the bureaucracy and the military—often act as powerful autonomous (that is, not controlled by the main three branches) actors, evidencing strong ability to garner resources for themselves at public expense and to thwart new policy initiatives.

Often—indeed, perhaps more often than not in most countries—the interests of these various state constituencies do not run in the same direction. The executive may try to emasculate the legislature and control the judiciary. The bureaucracy may become more concerned with seeking rents than with implementing the policy directives of the executive. The military may vie with the bureaucracy in rent seeking while evading the executive's attempts to assert constitutional control over it. All these disorders are common in the developing world (and many of them threaten to emerge in the industrial world as well at times).

Civil society and civil society organizations. Many constituencies can be said to be part of "civil society," long a highly contentious term claimed by various partisans for different purposes.[4] In the development community, some version of Gordon White's definition has increasingly become accepted. This definition holds that civil society consists of "an intermediate associational realm between state and family populated by organisations which are separate from the state, enjoy autonomy in relation to the state and are formed voluntarily by members of society to protect or extend their interests or values" (White 1994: 379).[5] Civil society can also be thought of as a third (nonprofit) sector of organized society, to be distinguished from the public (state) and private (business) sectors.[6]

An actual civil society constituency generally has some representation in the political process through one or more CSOs that advocate on its behalf.[7] Sometimes constituencies act directly in the political arena, as in voting or rallies and demonstrations (and even spontaneous or orchestrated riots); usually they are represented by CSOs.

Many CSOs are formal organizations, such as a mine owners association or a bus drivers union. Others are informal and ad hoc, such as a group of junior military officers claiming to represent a large sector of the army. Some exist outside the state sector altogether (mine owners), while others may have quasi-state status (junior army officers or public sector bus drivers). The key factor is whether a CSO is acting independently of the state.

Often there is contention over who is actually representing whom. There may be rival mining associations, splits within the bus drivers group, or contention between different groups of military officers. The boundaries between *constituency* and *civil society* can thus be murky. It is important to note that the term *constituency* denotes both a group of people able to make some claim on the policy process and representative advocates in the form of CSO(s) working on their behalf.

Some CSOs lack direct or obvious member constituencies. Human rights organizations often act on behalf of people they do not know; think tanks may act in the name of a wider public interest, such as protecting the environment.

The media are the hardest to fit into any civil society taxonomy, because they usually do not act as advocates for a particular constituency. Their function is to develop information and provide it to civil society. The media are critical to the democratic process; without free media to investigate and publicize state malfeasance—and to report the political scene more generally—democracy cannot exist for any length of time.

Constituency lifetimes. Constituencies can be short term or long term, temporary or sustainable. Short-term constituencies tend to coalesce in response to specific crises (crime in the neighborhood, point-source pollution of a water resource). While the CSOs representing them may enjoy some success in ameliorating the immediate problem, they generally do not endure long thereafter, and no constituency remains in any coherent form.

Long-term constituencies have enduring interests and CSOs with organizational skills, resources, and commitment. Examples are business associations, agricultural commodity lobbies, trade unions, and environmental organizations.

Participation

Participation constitutes the flip side of accountability. It refers to the processes through which constituencies (usually but not always through CSOs) make claims on the policy process. Voting, lobbying, lawsuits, bribery, demonstrations, letters to the editor, and petition drives all constitute modes of participation. Without participation, constituencies cannot demand accountability from the state.[8]

Strength of the State

A "strong state" is one that can enforce its rules of the political game in sectors such as the rule of law, freedom of speech, the right to vote, and minority rights. Accordingly, as used in this chapter, *strength* is not related to autocracy or to the degree of centralization the state attempts. Autocratic and highly centralized states may well prove unable to enforce their rules, as is the case in many African countries, while democratic and decentralized states can be very strong (as, for instance, in Canada).

Causal Linkages

This section examines how the concepts and variables introduced above can be linked to depict the policy-making process.

Policy Making and Linearity

In the abstract, policy making is a linear process moving logically from initial assessment through formulation and implementation, with a feedback system provided by monitoring and evaluation to effect improvements (figure 6.1). The presence of constituencies making demands throughout all phases, however, makes the process distinctly nonlinear (figure 6.2).

The abstract policy process depicted in figure 6.1 becomes a small activity within the hexagon at the center of figure 6.2. Constituencies and their representatives are constantly trying to impede, change, embellish, and add to policies or potential policies, forcing policy formulators to move back to the design stage to include new considerations before being able to proceed. An environmental activist group, for example, could stall a dam project already approved by the executive, forcing a new environmental impact analysis that takes potentially displaced

FIGURE 6.1
Policy Making as a Linear Process

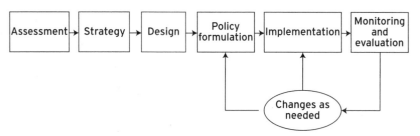

Source: Author.

FIGURE 6.2
The Policy-Making Process as It Actually Works

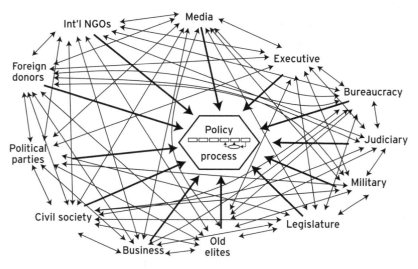

Source: Author.

floodplain dwellers into account. A coalition of industries might then pressure the legislature to enhance the preferential rates promised for hydropower generation from the dam. Civil engineers from the military might see an opening for involving themselves in flood control, which had been the preserve of the ministry of irrigation, and lobby the legislature to give them a role. Embarrassing newspaper coverage might reveal that the minister of public works had accepted a bribe to award contracts to cronies.

In one sense, all these interventions interfere with the policy-making process, degrading what should be a smooth flow of technical expertise leading from concept to execution. Indeed, they have often been perceived as such within the donor community. But in another, more important way, the seeming mess depicted in figure 6.2 reflects contributions to the policy-making process. The various state actors are shown as the nearly autonomous players they so often are, and international actors are included as well. All players endeavor to influence policy making at all stages; they also try to influence one another (thus the thicket of causal arrows). To the extent that they have the capacity and freedom to do so, the media and think tanks try to discern and report on these attempts at influence.

The bureaucracy may interact with the executive and the legislature to draw up policy initiatives while seeking to enhance its own perquisites as a policy is drafted. It may work with old elites bent on preserving their control of land or inefficient protected industries, as well as with business elements wanting to encourage investment that will displace those old industries.[9] It may also work with CSOs trying to improve the position of their constituents, whose interests (such as higher wages or primary education) likely conflict with those of older elites. Alternatively, the bureaucracy might side with a military wanting to keep wages down in order to enhance its own recruitment or wishing to divert education funds to its own activities. And on and on it goes.[10]

Engines of Environmental Accountability

For environmental accountability to be exacted, there must be institutions that demand it. Over time political parties can come to see the value of promoting policy initiatives that benefit the environment, as they begin to perceive constituencies concerned about environmental issues; this has been a frequent pattern in developed countries. The state itself can initiate the process (as the Indonesia case discussed below suggests). The media are always an essential instrument, for it is through them that constituencies can be informed, aroused, and mobilized to work through CSOs, parties, or the courts to seek environmental redress.

CSOs are the primary engine driving environmental accountability, doing so on behalf of constituencies that demand it. Sometimes they do so by advocacy directly targeting the state, sometimes they do so by appealing to political parties. It is the multiplicity of avenues that gives rise to the knotty and complex web portrayed in figure 6.2. In short, the engines are many and the paths are tangled.

Externalities and Constituencies

Public policy decisions often benefit certain constituencies while imposing costs on others. Allowing a mine to discharge its heavy-metal residues into an adjoining lake may kill or poison aquatic life, destroying local fishing industry and injuring the health of local fish consumers. Permitting loggers to clear-cut a tropical forest will degrade the soil, increase runoff, and cause downstream flooding and siltation. Because long-term constituencies such as mine operators and loggers tend to be relatively wealthy, they are able to impose negative externalities on the environment and society in order to direct resources toward themselves. In contrast, groups addressing externalities tend to be more ad hoc, episodic, and less well-endowed with resources. A central challenge to the development community is how to build and nurture long-term constituencies that can begin to redress externalities and redirect their benefits.

The Danger from Elites

Just as economic elites continuously try to undermine a market economy by creating oligopolies and oligopsonies, political elites constantly endeavor to attain dominance over policy making by establishing collusion among their own ranks. These tendencies manifest themselves just as much in advanced industrial countries as in the least developed ones. A strong state (in the sense defined above) is needed to protect against both perversions—to keep traditional elites (such as landowners and the political class in Latin America or new oligarchs in the former communist countries) from dominating while encouraging smaller businesses to operate in the economic game and less powerful constituencies to participate in the political game. In environmental terms this usually comes down to the bigger developers—miners, loggers, ranchers, fishers—having more money, better organization, greater access, in short more economic and political clout than those favoring environmental protection. But developers do not always win, especially if strong pro-environmental constituencies can be built to withstand them, ensure sufficient transparency to discover and disclose what goes on, and use appropriate institutions within the political system to demand accountability from the state for protecting the natural resource endowment.

Linking the Key Variables

Now that the main definitions and concepts have been established, it is possible to link the key variables. The linkage can be defined as follows: State institutions and potential polluters must be accountable in law and in practice to legitimate standards and nonstate constituencies, but accountability cannot be real without transparency and participation that extends well beyond elite circles.[11]

Accountability is the key concept in crafting effective environmental protection efforts. The central task is to promote the development of constituencies representing those not shown as entrenched players in figure 6.2, so that they can demand some accountability from policy makers amid the cacophony of interests and influences. For them to accomplish this task, transparency will be critical. The principal risk is that virtually any structural arrangements facilitating involvement of pro-environmental interests in the policy process also enables those who are better-off and already more advantaged to enhance their own positions.

Case Studies

The two case studies presented here describe successful environmental activism. In the first, two determined environmental CSOs mounted efforts that over time lead to effective initiatives to reduce atmospheric pollution in Delhi. In the second, an Indonesian government agency crafted an initiative to induce factories to reduce their emissions.

Reducing Air Pollution in Delhi

Sparked by two environmental CSOs, the city of Delhi—certified in the mid-1990s as one of the world's most polluted cities—managed to reduce air pollution levels dramatically by 2004.[12,13] Although atmospheric pollution levels still exceeded international standards, carbon monoxide emissions had fallen 32 percent, and sulfur dioxide levels had fallen 39 percent (http://cities.expressindia.com/full story.php?newsid=85665); sulfur dioxide declined another 63 percent, an even more dramatic improvement (Time Asia 2004). ·

Most of these reductions were brought about by cutting vehicular emissions, which had been widely believed to account for some 70 percent of air pollution. Laws designed to control atmospheric contaminants had been on the books since the early 1980s, but, as in so many developing countries, enforcement had been essentially nonexistent. At the beginning of the new millennium, however, the situation was beginning to change.

In a sense, the solution was simplicity itself: two CSOs—the Indian Council for Enviro-Legal Action, led by M.C. Mehta, and the Centre for Science and Environment (CSE), led by Anil Agarwal—brought public interest lawsuits and generated a high-profile, fact-based, publicity campaign that compelled the government to enforce the legal requirements. But the story behind this drama is much more complex and provides a first-rate example of how civil society activism works, what is needed to enable it to function, and how it fits into the wider context of accountability in governance.

Critical success factors. For the key actors in the drama—the Supreme Court of India, civil society, and the media—to have succeeded in imposing new environmental standards, several necessary conditions had to be in place, especially in the legal sector. The Supreme Court had to be autonomous enough from the executive that it could render judgments that were not subject to governmental control. In addition, public interest law had to be an acceptable component of the legal regime—that is, private citizens had to have the right to bring suit against the state on the grounds that it was not implementing its own laws. Finally, the Supreme Court had to have attained sufficient public esteem to have the legitimacy to issue orders to the executive that the executive was obliged to implement. The government could have evaded the Supreme Court's orders at several points in the course of the drama—by declaring polluting diesel to be an officially "clean" fuel, for instance—but it chose not to do so.

The second key group of actors was the CSOs. For these organizations to operate, freedom of speech and inquiry had to be in place as an enforceable legal norm. Organizations needed to be able to operate free of government harassment or intimidation, so that they could conduct their investigations, publicize their findings, and bring legal actions in the courts. They also required leadership with sufficient dedication, perseverance, and resources to remain active on the air pollution front over the long haul. M. C. Mehta pursued his first cause—banning industrial effluents that had been eroding the soft marble exterior of the Taj Mahal—for a decade before attaining victory in the Supreme Court in 1993. And though he launched the air pollution suit in the early 1990s, it was not until 1998 that the Supreme Court issued its first comprehensive mandate for eliminating pollutants. Anil Agarwal established the CSE in 1980; by the 1990s it had the technical expertise to investigate pollution levels and the institutional know-how to disseminate their findings effectively.[14] Neither organization was easily dissuaded.

The third actor, the media, needed constitutional guarantees of freedom of speech, which, except for the brief period of Indira Gandhi's "Emergency" rule in the mid-1970s, they have enjoyed virtually without interruption since independence. They also needed a source of competently researched and understandable findings to disseminate to their readership.

Taking advantage of public interest litigation provisions, M. C. Mehta asked the Supreme Court to compel the Delhi government to enforce the clean air laws that had been passed some 15 years earlier. Responding to the suit—in the course of which the government was made aware of the scientific evidence made available by CSE—and conscious of the public awareness created through media dissemination of CSE's findings, the Court created a monitoring committee, the Environment Pollution Prevention and Control Authority (EPCA), which it empowered to make policy recommendations and to which it appointed the CSE as a member. The CSE was thus in a position to make its expertise available to the Court on what amounted to an insider basis while at the same time disseminating its views to the wider public through the media. Groups opposed to the measures recommended by EPCA—the automobile industry, bus operators, taxi and auto rickshaw drivers, and eventually commuters inconvenienced by strikes and Court-mandated sidelining of buses—were able to take their case to the Court and the public as well, although in the end they were not able to affect the outcome.

Taking heed of the recommendations given it by EPCA, the Supreme Court in effect took charge of the antipollution effort, directing the Delhi government to phase out leaded gasoline, eliminate public transport vehicles more than 15 years old (the worst offenders), mandate the use of premixed fuel for two-stroke engines (which powered some 70 percent of Delhi's vehicles), and require the use of compressed natural gas (CNG) in all Delhi buses. Fortunately, the Court had the legitimacy to compel the government to carry out these mandates.

Each of the three actors—the Supreme Court, the CSOs, and the media—had a critical role to play. Because they were able to act together, civil society was able to hold the state accountable in a way that elections by themselves never would have been able to do (because air pollution, though vital to public health, could never have become a salient enough issue by itself to determine an election outcome).

Cautionary notes. As with any account of events in the public policy arena, certain caveats must be noted. First, the process took a very long time. The relevant laws were enacted in 1981, but even by the mid-1990s little had been done to implement them. Lawsuits filed in 1996 took another six years to yield intermediate results and culminate in the CNG mandate of 2002. In this case the Supreme Court and the two sparkplug CSOs stayed the course; attention spans for public policy issues are usually considerably shorter.[15]

Second, while think tank CSOs such as the two analyzed here clearly have a vital role to play in holding the state accountable, they represent at most a very narrow base of public opinion. Both were self-appointed guardians of the public interest, without any broad mandate; neither had a mass grassroots movement supporting it. M. C. Mehta's CSO really amounted to a one-man show with a supporting cast. While the CSE was a real organization in its own right, it was nonetheless a small elite group founded and managed by a charismatic leader, largely dependent on grants from outside sources, and without a membership base.[16] The wider public was at best informed of the efforts undertaken by these CSOs rather than involved in formulating them.

Third, while the two CSOs emerged triumphant in this story, they were not the only civil society players involved. An open civil society means that opponents of any initiative, not just its supporters, are free to play as well. Automakers, bus operators, and at one point even commuters entered the fray in opposition to the antipollution campaign. This time around, they lost, but in other controversies opponents of environmental controls often win, sometimes even rolling back earlier victories.

Fourth, the air pollution control initiative was fortunate in that while its first few policy forays faltered, it proved possible to find solutions that were technically and logistically feasible. Standards could be set for various pollutants (such as sulfur dioxide, nitrogen dioxide, or suspended particulate matter), but equipment often fails to detect them or can be manipulated to return false negatives (allowing polluting vehicles to pass the test); test inspectors can easily be bribed. Most alarmingly, if diesel were prohibited or restricted or subjected to price increases intended to discourage its use, bus operators could easily switch to subsidized kerosene, which is just as noxious in terms of effluent.[17] More-workable solutions emerged in the form of selling premixed fuel for two-stroke engines (which actually saved the operators money on repairs), phasing out leaded gasoline, retiring buses older than 15 years, and mandating the use of CNG in public buses. The first measure found favor with end users; the others were relatively easy to monitor and enforce. Such solutions would be considerably harder to craft in other environmental situations, however. No-fishing zones and logging restrictions, for example, would be considerably harder to implement.

Fifth, the Supreme Court wound up not only demanding and monitoring government compliance with environmental laws but actually managing compliance—in effect, micromanaging it. One has to ask whether such behavior is the proper province of a judiciary or should best be left to the executive, even if the executive drags its feet and delivers less than a perfect product. Constitutionally, after all, it is the executive that is charged with implementing the laws, not the courts. One also has to wonder whether the Court leaned too heavily on CSE for advice. Even granting CSE's impressive level of technical expertise, it would probably have been better to seek a multiplicity of views.[18]

Sixth, political parties were absent. As a "union territory" in the Indian system, Delhi has an elected government, with political parties regularly contesting control. Ideally, issues of significant salience are taken up by the parties, as part of their efforts to attract votes. But while air pollution may have been gripping to the elite readers of the *Times of India* and other newspapers, it was not absorbing enough to become a major platform plank for either of the two major Delhi parties in power over the period under analysis here.[19] Civil society therefore had to take on the entire burden of environmental activism, a role that in other countries might be shared with a green party or even a major party.[20]

Finally, even though the antipollution effort secured major successes, metropolitan Delhi's air can scarcely be called clean. Buses may be running on CNG and two-stroke vehicles on premixed fuel, but the vehicle population is expanding so rapidly that these gains are being overrun by new threats.[21] Data collected by CSE in May 2005, for example, show that suspended particulate matter, respirable suspended particulates, and nitrogen dioxide exceed international maximum permissible standards by a factor of at least two and in some cases much more.

It can also be argued that parallel efforts to reduce industrial pollution (mainly by moving factories away from urban areas) had a greater impact on improving air quality than lowering vehicular emissions (World Bank 2005). Other Indian cities face equally serious problems. To claim, then, as the Indian secretary of environment and forests does, that the country has reversed the environmental Kuznets curve may be premature.[22]

In sum, the achievements recorded in this episode are but way stations on a very long path to attain a cleaner environment in Delhi. They nevertheless clearly show that when the right enabling institutional structure is in place with respect to the legal system, civil society, and the media, it is possible to make significant headway.

Using Public Disclosure to Abate Pollution in Indonesia

Against a backdrop of rapid Indonesian economic growth in the 1980s and early 1990s (when industrial growth often topped 10 percent a year), rising pollution came to be viewed as a major concern. The state responded with various semi-voluntary and largely ineffective programs. Then, in the mid-1990s, the government introduced a pilot scheme in which industries were to self-report their levels of water pollution. This Program for Pollution Control, Evaluation and Rating (PROPER) led to significant reductions in pollution levels before being suspended in the wake of the financial crisis during the late 1990s. After economic recovery (and a democratic transition), a much-enhanced PROPER II was introduced in 2001.

Analysis of PROPER offers many contrasts as well as some interesting similarities with the Delhi air pollution saga.[23] The program kicked off in early 1995, when the government's Environmental Impact and Management Agency (BAPEDAL) rated some 187 factories in several river basins in Java, Kalimantan, and Sumatra. The river basins were selected mainly because they participated in a largely voluntary program begun in the late 1980s.[24] The factories were asked to submit data on water pollution in terms of biochemical oxygen demand (BOD) and chemical oxygen demand (COD). The data were analyzed, checked where discrepancies were noted, and formulated into a five-tier color-coded ranking system with two failing (black and red) and three passing (blue, green, and gold) grades, as follows:

- Black: No effort to control pollution
- Red: Some effort to control pollution, but results remain below national standard
- Blue: National standard met in all measures (not just an average among measures)
- Green: Pollution at least 50 percent lower than national standard in every measure; basin has proper sludge disposal, good records, and adequate wastewater treatment system
- Gold: Demonstrated adherence to international standards for water pollution, air pollution, and hazardous waste.

In June 1995, the five factories meeting the green standard were publicly lauded by the vice president (no factory met the gold standard). All factories privately received their detailed results, with the understanding that a retest and public release of results would occur by December. It was the public release of the findings that gave the program its impact, bestowing honor or shame on the factories rated (as well as added or reduced incentive to potential investors). The program gradually expanded, with the number of factories in the program reaching 324 by mid-1998.[25] During 1995–96, roughly 100 factories submitted self-reports during any given month. This figure rose to more than 170 in the following two years. In addition to encouraging factories to submit self-reports, BAPEDAL also inspected them on a regular basis. During 1995–96 about 200 official inspections were conducted each year.[26]

In late 1997, the Indonesian economy was hit by the Asian financial crisis. While PROPER I endured for another year or so, with only slightly faltering ratings despite a rapidly failing economy, it went into a hibernation phase for the next several years. By July 2001 the now democratic government reactivated the program under the Ministry of the Environment (KLH), beginning with 85 factories in 2002 and expanding to 466 by 2005.

In its reincarnation, PROPER II took on a new look in several dimensions. This time inclusion was mandatory rather than voluntary, and ratings targeted not only water pollution but also air pollution, hazardous waste, and community relations. As before, a multitier review structure and initial private disclosure to individual factories preceded public release of the findings. Of equal importance was the introduction of a sweeping decentralization initiative undertaken by the government in 1999 (fully effective in 2001), which devolved significant authority and resources to the provincial and especially the municipal (*kota*) and regional (*kabupaten*) level. Among the sectors transferred to the local level was the environment, meaning that local elected councils (*Dewan Perwakilan Rakyat Daerah,* DPRD) became responsible for monitoring pollution and enforcing standards.[27] Enforcement mechanisms were expanded, with judicial prosecution added to public disclosure and embarrassment.

Not surprisingly, given the more ambitious standards, there were more egregious failures; 33 of the 85 factories (39 percent) received black ratings in the first round and another 35 percent were rated red, while just 14 (16 percent) were rated blue and only 8 (9 percent) were rated green. As before, however, factories upgraded themselves: by 2004 just 3 of the original 85 factories had black rankings, while the number of factories with green ratings swelled to 51. As the number of factories in the program expanded to 466 in the 2005 round, the ratings improved, with only 72 factories (15 percent) rating black and 221 (47 percent) rating blue. The green level continued to remain elusive, however, with only 23 factories (5 percent) receiving a green rating.[28] Press coverage has been ample,[29] and CSOs have become involved at the national level, denouncing PROPER for being too strict or too lenient with polluters.[30]

The PROPER II process is a complex one (figure 6.3). It consists of eight main steps:

Step 1. First data gathering. Factories gather monthly data on air, water, and toxic wastes and report to KLH.

Step 2. Second data gathering. KLH develops database and analyzes the data.

Step 3. Data verification. KLH gives pollution data to each factory assessed in January–February, advising that it will measure again in several months.

FIGURE 6.3
PROPER II Annual Cycle

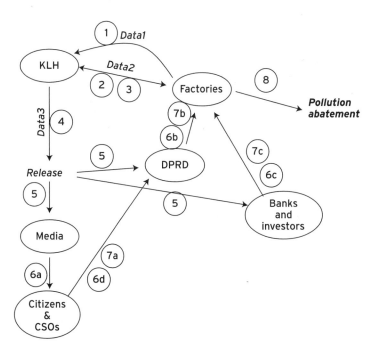

Source: Author.
Note: Numbers correspond to steps in to process (see text). KLH = Kementerian Negara Lingkungan Hidup.

Factories respond. With advice (if sought) from KLH, factories clean up their effluents and wastes (or fail to do so).

Step 4. Third data gathering. KLH measures pollution during the spring; ratings are finalized by advisory board (which includes Health Department representatives, business groups, and NGOs).

Step 5. Public data disclosure. Findings are made public in the summer in an official release, with details made available on the KLH Web site (http://www.menlh.go.id).

Step 6. Constituencies deliberate.

Step 6a. The media access the findings, decide how to report the story.

Step 6b. The DPRD (either *kota* or *kabupaten*) receives the findings, decides what action to take.

Step 6c. Banks and investors obtain and analyze the findings, decide how to respond in terms of investment policy.

Step 6d. Citizens and NGOs decide how to react.

Step 7. Constituencies react.

Step 7a. Citizens (individuals) and CSOs (groups) lobby DPRD.

Step 7b. DPRD decides whether to ignore, admonish, or prosecute factories ranked black or red.

Step 7c. Banks and investors decide whether to redirect loans and investments.

Step 8. Factories take (or fail to take) action on reducing pollution.

Critical success factors. Several factors proved key to success. First, and arguably most important, PROPER had an energetic champion in a position to push it along and make it work. Nabiel Makarim spearheaded the initiation of PROPER I and served as its head until he was removed as the program went into its hibernation phase in 1999. He then resurfaced in 2001 as minister in charge of the new Ministry of the Environment (KLH) to begin PROPER II. He was able to shepherd both the ministry and PROPER II through the rapid changes of government in 2001–24. Makarim's successor has continued to promote a strong environmental effort at the ministry, threatening to prosecute the 14 firms that received black ratings for the second time in 2005 if they did not improve their performance within a month (*The Jakarta Post* 2005a).

Second, BAPEDAL, the environmental agency, decided at the outset to play from its principal weakness as a regulator. Evidently realizing that it could not do much enforcing of environmental regulations in the ethos of corruption, autocracy, and crony capitalism that characterized the Suharto era, BAPEDAL did not even attempt such an approach. Instead, building on an approach first tried with an earlier project (the PROKASIH project, in which Nabiel Makarim was a key player [Afsah, Laplante, and Makarim 1996]), it relied on a recipe of public disclosure and anticipated reaction. It cushioned its disclosures by first releasing its test results privately to each factory and then allowing a six-month period before a retest and public data release, in order to encourage polluters to clean up. On the factory side some companies were induced to improve their performance by the combination of potential bad publicity; increased pressure from the news media, CSOs, and nearby communities; and the negative influence on stock prices and potential investors. In addition, some managers saw good ratings as helpful to them in obtaining an official certification from the International Standards Organization, which would help with future exports. This strategy substituted with some success for the inability to prosecute or penalize environmental code violators.[31]

Third, the combination of self-reports and government inspections proved sufficient to motivate a high level of honesty in the self-reporting process. In the first two years of PROPER II, only some factories were inspected, but the possibility of inspection evidently induced reasonably honest self-reports. By 2004, all factories included in the system were reporting self-inspection results monthly and being inspected at least once a year. These self-inspections (reported to the government) meant that factories could monitor their own progress. In a survey of factory managers involved in the PROPER catchment area, this feature of the program was selected as first or second most important by more managers than any of its other aspects (Afsah, Blackman, and Ratunanda 2004).

Fourth, the legal environment supported the program. Environmental Act 23 of 1997 established a requirement to monitor environmental compliance with regulations. It required companies to release environmental information and created legal rights to environmental information and rights for communities to participate in environmental management activities. That the act was passed during the Suharto period most likely indicates that it was intended as public relations window dressing. In the succeeding democratic era, however, its provisions are presumably enforceable and justiciable, giving PROPER II significant standing as an environmental initiative while providing civil society and the media the legal underpinning they need to promote pollution abatement. The media are free to criticize PROPER, and CSOs can chastise the government when they perceive that it is insufficiently dedicated to environmental goals.

Fifth, PROPER tailors the data it provides to its audiences, providing complex and specific data for factory managers and environmental experts while disseminating simple and straightforward data for the public and potential investors. PROPER II provides more than 70 indicators for gauging the pollution dimensions it tracks; the five-color coding scheme facilitates an instant understanding by the average citizen. A computerized data analysis system enables quick dissemination of results that would have taken more than a year to assemble using older spreadsheet methods.

Sixth, the existence of a sister environmental program provides synergy between the two efforts that surely benefits PROPER. The Good Environmental Governance program (known in Bahasa as the Bangun Praja Program [BPP]) began in 2002 as a voluntary initiative through which KLH would monitor municipal water quality and liquid and solid waste management. BPP operates in a fashion somewhat similar to PROPER, in that KLH assesses performance and makes public its findings in order to encourage local government responsiveness to citizens and citizen participation in local governance. By the program's third year, some 133 DPRDs had signed on to the program (Arundhati 2003; Leitmann and Dore 2005).

Finally, the provision of accurate, up-to-date information on pollution means that recipients can respond in a timely fashion while the data are still fresh. Factory managers can take corrective action, community residents can check claims for improvement against their own experience from living next to the factories, and potential investors can better determine where to put their money. The banking industry, for example, declared publicly in 2005 that it would not provide loans to companies included on the black list (Bisnis Indonesia 2005).

Cautionary notes. While PROPER has developed an ingenious strategy and achieved some notable successes, there are a number of reasons to be cautious in assessing the program. First, although it now covers many of Indonesia's largest factories and plans to expand to the entire country, PROPER still has a long way to go before it achieves complete coverage. Indonesia reportedly had some 20,000 factories in the mid-1990s; that number has surely expanded considerably since then. The 466 factories covered by PROPER II as of 2005 probably represent no more than 2 percent of the total (though by taking on large factories, it is surely covering a good deal more than 2 percent of the country's aggregate pollution).

What are the chances that PROPER II can scale up to anything like complete coverage? It could try to follow the path intended for PROPER I, which planned to expand by 2000 to cover the worst 10 percent of the country's factories, which contributed 90 percent of total water pollution (Wheeler 2000). With its much more ambitious attempt to monitor air pollutants and toxic wastes, can PROPER II be expected to attain this level of coverage within any reasonable time frame? To do so would be asking a great deal even in a highly developed country.

Second, and closely allied to the first point, is the issue of local government competence. The environmental laws of 1999 assign responsibility for environmental management to the DPRD. How competent are these councils to discharge their duties in this regard? How will they respond to the pressures that are certain to be brought to bear on them from civil society, the national government, and factories? Given that the *kotas* and *kabupatens* had virtually no environmental responsibilities before 1999 (and acquired very little experience before the beginning of PROPER and its sister programs in 2002), the capacity of DPRD staff, especially its elected council members, to deal with environmental issues will be a major challenge for some time to come.

The KLH can support the learning process on the technical side; the political side will likely prove a good deal harder to master, for the council will have to choose between environmental protection and development, trying to accommodate and encourage both of these often conflicting goals without compromising either. This is a difficult balancing act, as the World Bank (2001, 2003) has recognized. Devolution of governance power can facilitate responsiveness to local concerns, allow local voters to hold government accountable, encourage more sustainable use of resources, and so forth. But it can also facilitate local elites taking control, steer benefits to themselves at the expense of the general public, and covering up their misdeeds. The track record in decentralization efforts, especially in developing countries, has been mixed at best, in the natural resources management sector as well as more generally. In other countries local government bodies have proven themselves able to hold polluters to account—but they have been just as likely to sell out the environment as to protect it.[32] Local CSOs, free and vigorous media, and institutions such as user groups can act as a prophylactic against such depredations, but their triumph is far from certain and always in danger.

A third and allied question regards how to craft and strengthen mechanisms that can build on PROPER to improve the environment in Indonesia. The media, civil society, and the banking sector have become engaged, which is no mean achievement. The media and civil society players spread and magnify unfavorable publicity ("naming and shaming"), putting moral and social pressure on errant firms to improve; the banking sector wields the usually more powerful tool of monetary clout. Will these tools be sufficient to the task? Surely they will be in some—perhaps many—cases. But if Indonesian industries are like those in the rest of the world, stringent government enforcement of norms will be necessary to significantly reduce pollution.

The DPRD, which is institutionally charged with responsibility for the environment, has not yet entered the fray in any substantial way. What can a DPRD do to encourage firms to improve their environmental performance? Can it set standards and enforce them by withdrawing licenses, levying fines, or even shutting down flagrant violators? The KLH has threatened to prosecute factories rated black two times running that have not taken action. But local governments do not appear to have acted on this front, and the issue of jurisdiction (can the central government bring a legal case in an area reserved by law to local government?) is not clear. How this will play out over time will be critical.

Fourth, there is the issue of corruption. The Suharto regime was notoriously corrupt. Although successor democratic administrations have improved, corruption remains profound at all levels. Given Indonesia's distinctly unenviable track record in sectors such as logging and mining, it would be surprising if anything more than the most gradual rate of expansion in PROPER's coverage were not accompanied by serious levels of corruption.[33] Self-reports can be faked, and inspectors can be bribed, threatened, or both.

Fifth, while the five-color scheme is a brilliant one in many ways, providing easily assimilable information, it masks a great deal of very useful information. The key category is the red rating, which indicates that although some effort is being made to control pollution, performance falls short of the national standard. This category covers a huge range of performance. The BOD standard for plywood factories under PROPER I, for example, was 100 milligrams per liter of discharge; any plant generating more than 100 milligrams would be coded red (assuming that it had taken at least some minimal effort to deal with pollution). Thus if a factory had reduced BOD effluent from, say, 312 milligrams (one standard deviation above the observed mean for all plywood factories) to 104 milligrams—a reduction of two-thirds—it would continue to be rated red (López, Sterner, and Afsah 2004). Similarly, a factory that at one rating period was on the verge of passing from blue to green status but then fell to just above the red level would continue to be rated blue, even though its performance had deteriorated badly. Moreover, the BOD rating itself simply measures the amount of pollutant per liter of discharge, not the total pollutant discharged from a given plant over a particular period of time. Accordingly, a factory could clean itself up significantly for reporting purposes just by increasing its water intake and discharge. The same issue affects air pollution and toxic waste disposal under PROPER II.

An allied cause for concern is the state's position on enforcement. The minister of the environment has declared his intent to prosecute firms failing to move from black to red status, but all a factory has to do to gain such an upgrading is to show some sign of effort, however feeble or even hypocritical.

Patterns and Themes

Both case studies reveal lessons about accountability, transparency, and the need for long-term constituencies to ensure accountability. They also illustrate the general untidiness of critical public decisions in a democracy. In each of the two cases, the three major themes of this chapter played out differently (table 6.1).

Accountability

In India, CSOs were able to take advantage of the opportunities offered by the legal environment to hold the state accountable for regulating the environment. In Indonesia, BAPEDAL realized that corruption and cronyism rendered such a course impossible. It therefore devised what might be termed a jiujitsu martial art tactic by turning the business community's longer-term need for new investment against its shorter-term need for quick profit through polluting the environment. Business (or at least a good part of it, for only some firms decreased their pollutants) substituted for the state in becoming accountable and taking ameliorative action. Business concern for enjoying a good reputation also played a role. In PROPER II these needs were again harnessed, supplemented by a government threat to prosecute laggard firms.

TABLE 6.1
Accountability, Transparency, and Long-Term Constituencies in Two Case Studies

Case Study	Institution Held Accountable	Machinery Providing Transparency	Long-Term Constituency Supporting Better Environment
Delhi air pollution	State	Legal environment	Civil society organizations, engaged citizenry
Indonesia PROPER	Industrial community	Self-reporting, inspection, public disclosure	Business elites, (gradually) civil society organizations

Source: Author.

While the successes in both countries were important, any celebration of them must be tempered by realistic assessment. Delhi's air pollution continues to far exceed maximum internationally permitted standards, and the explosion of new vehicles (cars and trucks) not subject to the regulations threatens to undo the progress made. In Indonesia PROPER II expanded of the list of environmental wastes it covers (air and toxic wastes are now included as well as water), and the number of factories covered rose more than fivefold (from 85 to 466 over a three-year period). Much needs to be done before the country's more than 20,000 factories are included, however, and the national standards (the blue level) remain significantly below generally acceptable international yardsticks. Even a cold dose of reality should not obscure the facts, however, that in both countries transparency was used to launch initiatives demanding accountability and the support of constituencies made their success possible.

Transparency

Transparency in Delhi came through the legal environment, which guaranteed freedom of speech and inquiry and permitted public interest lawsuits, thus allowing CSOs to pursue their advocacy programs. CSOs could investigate malfeasance, publicize their findings, and bring legal action against the state demanding that it enforce the environmental regulations on its books. All of these activities showed what the state was doing or failing to do.

Indonesia was also able to achieve a good degree of transparency, even during the Suharto era. As a government organization with an able and determined leader, BAPEDAL was able to gather pollution data, and the media were able to publish it. Bad publicity, combined with concerns for future investment, proved sufficient to effect some improvement (as seen in the higher rankings attained by many firms in successive ratings). CSOs had less room for advocacy maneuver than in India, and the courts were subservient to the wishes of the executive branch. PROPER I operated with a good degree of transparency, but other elements of the system made it difficult for civil society or the legal system to take advantage of the information disseminated to the press.[34] PROPER II presents far more opportunity.

Constituencies

The fundamental long-term constituency in India was the active citizenry—people who participated in civic life by reading newspapers, voting in elections, supporting the constitution's division of powers, and showing indignation and anger when the operating rules of the political order were flouted. Because these citizens respond in opinion polls, support candidates for office, and vote, political leaders found they must pay heed. In Indonesia, an active citizenry was important, for it was they who read the media's accounts of pollution and supported CSOs. They were reinforced by banks and potential investors, who had the power to decide whether or not to back the firms financially.

When it comes to nurturing these constituencies, elite investors can generally look after themselves, as they have done everywhere throughout most of modern history. Maintaining an active citizenry that will oppose environmental degradation and political corruption is more difficult. Many institutions are required to keep it in place and dynamic, among them a vibrant media, a vigorous public discourse, self-motivated CSOs, and strong civic education in the school system. High-profile support from national leadership can be a powerful support as well.[35]

Many players got into the act as each of the two cases unfolded. Of the institutions shown in figure 6.2, all but the military (and perhaps old elites) were involved directly or indirectly in the Delhi air pollution case (figure 6.4). The circle of involved players in the Indonesian example was somewhat smaller (figure 6.5), but over time it came to include all the Indian actors except the parties and the judiciary. Old elites (including major industrialists from the Suharto era) will necessarily be drawn in if they have not been already. If and when foot-dragging industries are prosecuted by the state, as the minister of the environment has promised, the judiciary will get involved, and it surely cannot be too long before at least some political party finds some aspect of pollution abatement sufficiently appealing to take on board in its efforts to appeal to voters. It is also likely that the military

FIGURE 6.4
Environmental Policy-Making Process for Air Pollution Abatement in Delhi

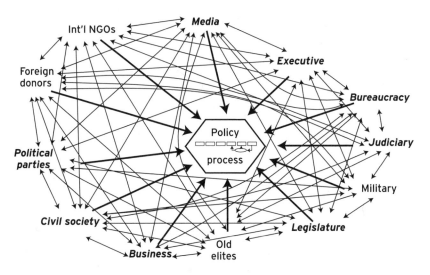

Source: Author.
Note: Significant players shown in boldface italics.

FIGURE 6.5
Environmental Policy-Making Process Under Indonesia's PROPER II Program

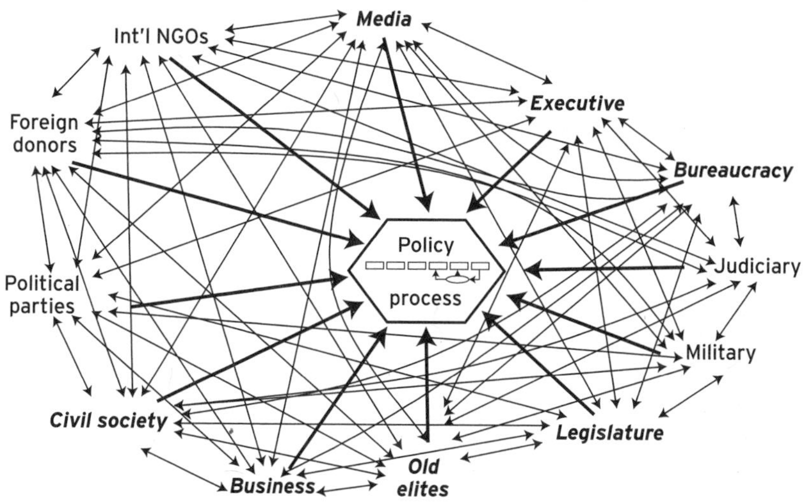

Source: Author.
Note: Significant players shown in boldface italics.

may well have played some shadow role, operating as it does a panoply of business enterprises, including many industrial ones.[36]

The role of foreign donors and international NGOs was somewhat obscure in both cases. The CSE in Delhi, for example, has received grants from the Ford Foundation, and it maintains links to international environmental NGOs for exchanging information and experience. WALHI has enjoyed support from a number of outside organizations, such as Friends of the Earth, Greenpeace, and the Dutch NGO Novib.[37] Both phases of the PROPER initiative received support from the World Bank.

The Role of Civil Society Organizations

CSOs served as catalysts for the Delhi case, but they did not carry the load of exacting accountability alone. Instead, they initiated processes that enabled other institutions to demand accountability. Environmental activists in Delhi convinced the judiciary to enforce what the municipal authorities had been required to do by law all along. In Indonesia CSOs had at best minimal involvement during the Suharto era; during PROPER II they have been active on the environmental front,[38] although they have not assumed the central role of their counterparts in India.

Despite their foreign linkages, the Indian CSOs analyzed here have long been pursuing self-generated activities. Likewise, although it has long had foreign connections, WALHI has largely determined its own agendas in the environmental field. Unlike some NGOs, especially in the first flush of grant largesse that often comes when donors move into new countries, these were not "briefcase NGOs" or "family NGOs" pitching their program agendas to chase foreign funds. Both Indian CSO efforts drew on long experience. The CSE began its work in the early 1980s; Mehta's organization launched its first campaign at about the same time. WALHI dates back to 1980, when it began as a coalition of 10 environmental NGOs. By the time the issues covered here were taken up, both organizations had considerable experience at public interest initiatives.

CSO internal accountability and transparency per se are not as important as the accountability and transparency they can demand from the political system. Among the CSOs in the case studies, M. C. Mehta's enterprise seems largely a one-man operation, while CSE has been a group operation created and nurtured by a charismatic leader. Among successful CSOs, strong, even autocratic leadership tends to be the norm, as might be expected in a milieu in which achievement tends to depend more on gritty determination and even obsession than on anything else. It should not, therefore, be surprising that many CSOs are less than fully democratic in their internal operations. These organizations are more important for what they do than what they are, however: their worth lies much more in their ability to hold the state to account and to represent their constituency than in their capacity to serve as models for democratic internal management.

The Legal Environment

CSOs in India were able to launch their initiatives by acting within an enforceable legal environment that sanctioned their activities. Freedom of speech and inquiry were firmly in place, though they had been seriously abridged in the not too distant past with Indira Gandhi's "Emergency" in the mid-1970s. After the restoration of democracy, CSOs could investigate malfeasance and the media could report it. Transparency had become a part of the institutional landscape.

Indonesia created a suitable legal environment on paper with its 1997 Environmental Act, but during the Suharto era its provisions were barely enforced. It was a government agency, BAPEDAL, that superintended the self-reporting program and the release of findings. After Indonesia's democratic transition, the legal environment became much more supportive of free speech and inquiry, and CSOs such as WALHI have virtually all the legal room they need to advocate their cause.

By itself, however, transparency would not have been enough: the truth had to be not only discovered but also disseminated through the media. It is hard to overstress the importance of the media in this regard. It is not just that the truth must be known to some (a few always know the inside story) but rather that people generally (the public, investors, shareholders) must know it.

Within the constitutional structure, separation of powers had established itself sufficiently firmly in India that the Supreme Court enjoyed autonomy from executive control. In Suharto's Indonesia, all branches of government remained firmly under presidential control. The legislature amounted to a rubber stamp, meaning that the Environmental Act of 1997 was in fact a creation of the executive (in contrast with the decentralization legislation of 1999). Courts could enforce it only on permission of the executive.[39] Fortunately, BADEPAL proved able to devise a system that did not depend on enforcement power from the state but on public dissemination to create pressure for abating pollution.

The Process of Public Decision Making

The underlying theme in the two country case studies is the exaction of accountability from the executive. In neither case was the path to success straightforward. The processes were convoluted in the Delhi case and indirect in the Indonesian one. Indian officials found it difficult if not impossible to set enforceable vehicular air pollution standards and to restrict the use of polluting fuels such as diesel; eventually, more-workable solutions emerged that could be enforced. Indonesian officials devised a scheme under which a weak state seriously penetrated by cronyism and corruption never had to undertake direct enforcement but instead could rely on market forces and internal industry incentives to provide the incentive for pollution abatement.

In the Indian case, civil society advocacy achieved demonstrable results: vehicular air pollution in Delhi declined appreciably. In Indonesia the role of civil society was more subtle, but CSOs such as WALHI did play roles in reducing factory pollution significantly.

In many (probably most) settings, assessing outcomes is considerably more difficult, particularly when it is poor and vulnerable constituencies on whose behalf advocacy is undertaken. Gauging the effects of efforts to demand accountability should be a major concern for donors promoting civil society initiatives, as well as others (such as the Indian Supreme Court) that rely on them.[40]

Conclusion

Accountability, constituencies, and transparency are mutually dependent, in that each needs the other two to endure if public policy is to be inclusive and improve the environment on a consistent basis. Accountability will not mean anything without constituencies to exercise it, and constituencies cannot make informed decisions in demanding accountability without transparency. But accountability is the master concept; the other two serve to support it. For it is by holding the state and polluters accountable that externalities injurious to the environment can be reduced and corruption and malfeasance in high office opposed.

The two case studies differ significantly along all three dimensions. In Delhi, it is the state that is held accountable for enforcing environmental standards; in Indonesia it is the industrial community. In Delhi, the legal environment provides the critical requirements for transparency to function; in Indonesia, the system of monitoring and disclosure provides the essential mechanism. In both cases, pollution abatement depends on civil society as a long-term constituency for support; in Indonesia the self-interest of business investors and banks has also been harnessed.

Notes

1 Another way to put this might be in terms of "inclusive decision making," as analyzed in chapter 3 of this volume.

2 Examples of such times of systemic stress include Viktor Yushchenko's 2004 Orange Revolution in Ukraine, in which huge public demonstrations lasted weeks in Kiev (see Kuzio 2005), and the firestorm of protests that erupted in the United States after President Richard Nixon fired his attorney general at the height of the Watergate crisis in 1973. Mass movements or campaigns have taken place over environmental issues as well. Examples include the widespread (and sometimes violent) resistance to British attempts to enclose Himalayan forests in the late 19th century (Guha 1990) and the sizable groups currently mobilized for and against forest clearance in Amazonian Brazil.

3 *Organization* here does not necessarily mean formal institutionalization. It can mean a group pattern of behavior existing over time within an understood set of operating rules, a definition the Orange Revolution demonstrators fully met.

4 For a succinct discussion of the term, see Candland (2001). Ehrenberg (1999) traces the concept of civil society from Aristotle to Jürgen Habermas and Robert Putnam.

5 White's definition has become widely (though by no means universally) accepted in the highly contested realm of specifying the meaning of this term.

6 In this conception of the term, individual business firms are in the private sector, whereas an association representing the business community is part of civil society. For an exploration of civil society seen as the larger nonprofit sector, see Salamon, Sokolowski, and Associates (2004).

7 CSOs often deliver services. This chapter focuses on their advocacy function. For an analysis of how the two functions relate to each other, see Blair (2002).

8 *Participation* has had a longer history in the development community than any of the other terms used in this chapter, beginning with the perceived necessity to include the "felt needs" of villagers in the community development initiatives of the 1950s. It has had a considerably more checkered career than accountability, beginning with Albert Mayer's work in Indian community development in the 1940s (Mayer 1958) and Gunnar Myrdal's devastating critique in the 1960s (Myrdal 1968).

9 The dividing line between "old elites" and newer "business elements" is hazy at best. One challenge for environmental policy making is to convince old elites commanding polluting industry to turn themselves into modern entrepreneurs who attract new capital in part through operating environmentally sound enterprises.

10 Not all these interests are totally self-seeking. Many bureaucrats feel a duty to deliver the services they are supposed to deliver; the executive often believes itself to be on a mission to nurture economic growth; business interests often take pride in their products.

11 *Legitimate* here means standards enacted and constituencies behaving within parameters established through constitutionally determined procedures or operating rules of the game.

12 Except where noted, the facts of this account (but not their interpretation) are drawn largely from Bell and others (2004). See also World Bank (2005).

13 Delhi's claim to be the one of the most polluted cities in the world lies partly in its inclusion in the short list of about 20 large cities monitored and publicized by the World Health Organization. In fact, many cities in India have far worse pollution (Agarwal n.d.; UNESCAP 2000, cited in World Bank 2005). Delhi's pollution is certainly heavy, however, a fact that has become widely known and commented on by residents, visitors, and the media. This high level of public awareness undoubtedly made it easier to gain legal attention for the lawsuits that proved instrumental in curbing pollutant levels.

14 A LexusNexus search of the *Times of India* (Delhi's highest-circulation newspaper) and its sister publication the *Economic Times* (arguably the country's leading financial daily) yielded some 385 stories mentioning the CSE during the five years ending in May 2005— impressive evidence of the organization's ability to get its message out to the public.

15 In rare instances an environmental issue—such as the Narmada dam controversy in Western India—has engaged public attention for more than two decades.

16 The fact that CSE survived the death of its founder-leader, Anil Agarwal, in 2002 and continues to thrive indicates that it is much more than a one-man operation.

17 Kerosene subsidies would have been politically impossible to eliminate, because of the widespread use of kerosene for cooking and lighting among the poor throughout India.

18 CSE's inside position with EPCA may have given it undue influence. It is typically industrial groups that have the inside track in such matters.

19 The Bharatiya Janata Party governed Delhi until the 1998 election, when it lost to the Indian National Congress Party.

20 When in power, both the Bharatiya Janata Party and the Congress Party proved willing to follow Supreme Court mandates on air pollution. Politicians of both parties, even those who had opposed various aspects of the pollution control effort, took public credit for Delhi's cleaner air in the 1998 elections. Thus in the end parties did get involved.

21 The number of vehicles in Delhi rose from less than 250,000 in the mid-1970s to 3.7 million in 2003 and is expected to reach 6 million by 2011, with trucks and private passenger cars constituting a very large proportion of the increase—a vivid testimonial to the consequences of economic growth in India (*Times of India* 2004a, 2004b).

22 Nobel laureate Simon Kuznets posited that as a country develops economically, income inequality initially rises. The "environmental Kuznets curve" holds that environmental pollution increases and then decreases with per capita income along a similar inverted U-curve. Ghosh (2004) claims that India has begun to reverse the curve at a much earlier point (in per capita income terms) than international experience would have predicted.

23 Unless otherwise indicated, factual information about PROPER comes from Afsah and Dore (2005).

24 This was the PROKASIH program. For an analysis, see Afsah, Laplante, and Makarim (1996).

25 Data in this paragraph are from Wheeler (2000).

26 Data for PROPER I are from Wheeler (2000); Blackman, Afsah, and Ratunanda (2004); and López, Sterner, and Afsah (2004).

27 Each *kabupaten* and *kota* has a DPRD (the urban *kotas* exist independently of the largely rural *kubupatens*). The number of DPRDs has been expanding, from less than 300 when the decentralization law was passed in 1999 to more than 430 by 2003. The number of people living in a *kabupaten* or *kota* ranges from less than 25,000 to more than 4 million, with an average of roughly 500,000 (World Bank 2003). Among other provisions, the 1999 law more than doubled the subnational share of public expenditure.

28 For PROPER I and especially PROPER II, factories that were rerated in successive years improved markedly. Forty of the 112 plants rated black or red in June 1995 had progressed to blue by July 1997. There was also some retrogression, with 16 factories falling from blue to red over the same period (López, Sterner, and Afsah 2004). Under PROPER II of the 63 plants rated black or red in 2002, only 26 continued to be so rated by 2004.

29 PROPER findings were given a front-page story and a lead editorial in the leading English daily, *The Jakarta Post* (2005a, 2005b). The report was accorded even more extensive coverage in the country's most widely circulating Bahasa language daily, *Kompas* (2005a, 2005b). While the newspaper emphasized the worst performers, it listed all ratings on its Web site at http://www.menlh.go.id/proper/.

30 On the day following publication of PROPER's findings, the Federation of Indonesian Metalworking and Machine Industry Associations denounced PROPER for demanding higher standards than factories could afford (Hakim 2005); a few days later, one of the country's most prominent environmental CSOs, Wahana Lingkungan Hidup (WALHI), criticized KLH for awarding two green labels it thought were undeserved (WALHI 2005).

31 See Blackman, Afsah, and Ratunanda (2004) for an analysis of the PROPER I disclosure program, including a survey of factory participating factories.

32 For a more general analysis of the promise and problems, see Blair (2000) and Manor (1999). For an analysis of local governance and natural resource management, see Blair (1996). Indonesia's new decentralized governments have not proven immune to the lure of corruption (Borsuk 2003).

33 In its 2005 survey, Transparency International ranked Indonesia 137th out of 159 countries in its Corruption Perceptions Index. India ranked 88th that year (Transparency International 2005).

34 The regime in Indonesia during the 1990s might best be described as one of "soft authoritarianism"; restrictions on free speech, the media, and civil society were in place but not harsh (Sen and Hill 2000; Eldredge 2002).

35 On Indonesia's Earth Day in June 2005, President Susilo Bambang Yudhoyono delivered a widely publicized speech urging the public to get involved in campaigns to press governors, mayors, and regents to clean up their localities (Witular 2005).

36 In Indonesia, the military receives only a small fraction of its budget as an official government appropriation. It is expected to generate the vast bulk of its revenues from its own money-raising efforts, most of which involve business operations.

37 WALHI provides a list of donors on its Web site (http://www.walhi.or.id/pusinfo/).

38 A vigorous environmental activist initiative has also been directed at the U.S.–based Newmont Mining Company, the world's largest gold producer (Perlez and Rusli 2004; Perlez 2004, 2005).

39 In 2004, the judiciary officially became independent of the executive branch in Indonesia; it remains to be seen how effective this separation will be (U.S. Department of State 2005).

40 For more on measuring civil society advocacy outcomes, see Blair (2004) and Hirschmann (2002).

References

Afsah, Shakeb, Allan Blackman, and Damayanti Ratunanda. 2004. "How Do Public Disclosure Pollution Control Programs Work? Evidence from Indonesia." Discussion Paper 00–44 Resources for the Future, Washington, DC.

Afsah, Shakeb, and Giovanna Dore. 2005. "Instituting Environmental Programs in Developing Countries: The Case of Indonesia's PROPER Program." Draft background paper. World Bank, Environment and Social Development Sector Unit, East Asia and Pacific Region, Washington, DC.

Afsah, Shakeb, Benoit Laplante, and Nabiel Makarim. 1996. "Program-Based Pollution Control Management: The Indonesian PROKASIH." Policy Research Working Paper WPS 1602, World Bank, Washington, DC.

Agarwal, Anil. n.d. "When Will India Be Able to Control Pollution?" http://www.cseindia.org/html/au/anilji/airpollution_next.htm.

Arundhati, Sri Tantri. 2003. "Bangun Praja Program." Note for Second Meeting of the Kitakyushu Initiative Network (Mayors' Segment)." Weihai, China, October 17. http://www.iges.or.jp.

Bell, Ruth Greenspan, Kuldeep Mathur, Urvashi Narain, and David Simpson. 2004. "Clearing the Air: How Delhi Broke the Logjam on Air Quality Reforms." *Environment* 46 (2): 22–39.

Blackman, Allen, Shakeb Afsah, and Damayanti Ratunanda. 2004. "How Do Public Disclosure Pollution Control Programs Work? Evidence from Indonesia." *Human Ecology Review* 11 (3): 235–46.

Blair, Harry. 1996. "Democracy, Equity and Common Property Resource Management in the Indian Subcontinent." *Development and Change* 27 (3): 475–99.

———. 2000. "Participation and Accountability at the Periphery: Democratic Local Governance in Six Countries." *World Development* 28 (1): 21–39.

———. 2002. "Civil Society, Advocacy, Service Delivery, Non-Governmental Organizations, and USAID Assistance: A Mapping Exercise." Paper prepared for the Center for Development Information and Evaluation, U.S. Agency for International Development. Management Systems International, Washington, DC.

———. 2004. "Assessing Civil Society Impact for Donor-Assisted Democracy Programs: Using an Advocacy Scale in Indonesia and the Philippines." *Democratization* 11 (1): 77–103.

Borsuk, Richard. 2003. "In Indonesia, a New Twist on Spreading the Wealth: Decentralization of Power Multiplies Opportunities For Bribery, Corruption." *Wall Street Journal*, Eastern Edition, January 29.

Bisnis Indonesia. 2005. "Companies with Bad Environment Management Not to Get Loans from Banks." Bisnis Indonesia, Jakarta, June 24.

Candland, Christopher. 2001. "Civil Society." In *The Oxford Companion to Politics of the World*, ed. Joel Krieger, 140–41. New York: Oxford University Press.

Diamond, Larry. 1999. *Developing Democracy: Toward Consolidation*. Baltimore, MD: Johns Hopkins University Press.

Ehrenberg, John. 1999. *Civil Society: The Critical History of an Idea*. New York: New York University Press.

Eldredge, Philip J. 2002. *The Politics of Human Rights in Southeast Asia*. London: Routledge.

Ghosh, Prodipto. 2004. "Harmonizing Environmental Concerns and Economic Growth: The Indian Perspective." *Environment Matters* (Annual Review, July 2003–June 2004): 10–11.

Guha, Ramachandra. 1990. *The Unquiet Woods: Ecological Change and Peasant Resistance in the Himalaya*. Berkeley: University of California Press.

Hakim, Zakki P. 2005. "Govt Environmental Audit Improper, Untimely: Industry." *The Jakarta Post*, August 10.

Hirschmann, David. 2002. "'Implementing an Indicator': Operationalizing USAID's Advocacy Index in Zimbabwe." *Development in Practice* 12 (1): 20–32.

The Jakarta Post. 2005a. "Cost of Environment." Editorial, August 12.

———. 2005b. "Top Firms Get Poor Marks on Environment." August 9.

Kompas. 2005a. "Peringkat hitam bakal digugat." August 9.

———. 2005b. "Perusahaan hitam keluhkan proporsi." August 10.

Kuzio, Taras. 2005. "The Opposition's Road to Success: Ukraine's Orange Revolution." *Journal of Democracy* 16 (2): 117–30.

Leitmann, Joe, and Giovanna Dore. 2005. "Better Environmental Governance: Improving the Role of Local Governments and the Private Sector in Indonesia." Presentation prepared for the Environmentally and Socially Sustainable Development Week, World Bank, Washington, DC.

López, Jorge Garcia, Thomas Sterner, and Shakeb Afsah. 2004. "Public Disclosure of Industrial Pollution: The PROPER Approach for Indonesia?" Discussion Paper 04–34, Resources for the Future, Washington, DC.

Manor, James. 1999. *The Political Economy of Democratic Decentralization*. Washington, DC: World Bank.

Mayer, Albert. 1958. *Pilot Project India: The Story of Development at Etawah, Uttar Pradesh*. Berkeley: University of California Press.

Myrdal, Gunnar. 1968. *Asian Drama: An Inquiry into the Poverty of Nations*. New York: Pantheon.

Perlez, Jane. 2004. "Mining Giant Told It Put Toxic Vapors into Indonesia's Air." *New York Times*, December 22.

———. 2005. "Cause of Mystery Ills Splits Indonesian Village." *New York Times*, March 27.

Perlez, Jane, and Evelyn Rusli. 2004. "Spurred by Illness, Indonesians Lash Out at U.S. mining Giant." *New York Times*, September 8.

Salamon, Lester M., S. Wojciech Sokolowski, and Associates. 2004. *Global Civil Society: Dimensions of the Nonprofit Sector*, vol. 2. Bloomfield, CT: Kumarian Press.

Sen, Krishna, and David T. Hill. 2000. *Media, Culture and Politics in Indonesia*. Melbourne: Oxford University Press.

Time Asia. 2004. "Choking on Growth." December 13.

Times of India. 2004a. "Green Mirage: Delhi's Air is Murkier." September 23.

———. 2004b. "Inching towards a Greener Delhi?" June 7.

Tocqueville, Alexis de. 2000 [orig. 1835, 1840]. *Democracy in America*, trans. and eds. Harvey C. Mansfield and Delba Winthrop. Chicago: University of Chicago Press.

Transparency International. 2005. "Corruption Perceptions Index 2005." http://www.transparency.org.

U.S. Department of State, Bureau of Democracy, Human Rights, and Labor. 2005. "Indonesia." In *Country Reports on Human Rights Practices, 2004*. Washington, DC. http://www.state.gov/g/drl/rls/hrrpt/2004/41643.htm.

WALHI (Indonesian Forum for Environment). 2005. "WALHI and JATAM Question Green Label for Newmont and RAPP." Press release, August 12.

Wheeler, David. 2000. *Greening Industry: New Roles for Communities, Markets and Governments.* New York: Oxford University Press for the World Bank.

White, Gordon. 1994. "Civil Society, Democratization and Development (I): Clearing the Analytical Ground." *Democratization* 1 (3): 56–84.

Witular, Rendi A. 2005. "Susilo Wants Public to Shame Dirty Cities." *The Jakarta Post*, June 7.

World Bank. 2001. *Indonesia: Environment and Natural Resource Management in a Time of Transition.* Washington, DC: World Bank.

———. 2003. *Decentralizing Indonesia: A Regional Public Expenditure Review, Overview Report.* Report 26191-IND, World Bank, East Asia Poverty Reduction and Economic Management Unit, Washington, DC.

———. 2005. *For a Breath of Fresh Air: Ten Years of Progress and Challenges in Urban Air Quality Management in India, 1993–2002.* New Delhi: South Asia Region, Environment and Social Development Unit, World Bank.

CHAPTER 7

Learning in Environmental Policy Making and Implementation

Alnoor Ebrahim

THIS CHAPTER EXAMINES how "learning" occurs in the context of environmental policy formulation and implementation. It identifies ways in which learning can be enhanced to improve policy making and implementation.

The chapter begins with an overview of key types of learning apparent in organizational and environmental policy contexts. It then analyzes examples drawn from the experiences of Brazil and South Africa. It concludes by identifying three sets of factors affecting environmental policy learning: priority or agenda setting, stakeholder access and representation, and accountability.

Two cautionary notes are warranted. First, the approach presented here runs the risk of portraying learning as a rational and technocratic endeavor rather than a messy process embedded in social, political, and cultural ambiguities. Second, there is a risk of romanticizing learning (how can learning be anything but good?). In practice, learning processes are imperfect and do not easily lead to improvements in organizational behavior, suggesting a need for conservatism in expectations.

Alnoor Ebrahim is the Wyss Visiting Scholar at the Harvard Business School and a Visiting Associate Professor at the John F. Kennedy School of Government, Harvard University. He is grateful to Blake Kehler for his able research assistance. He would also like to thank Margaret Keck, Len Ortolano, and the volume editors for their comments and feedback, as well as David Bonbright, Faranak Miraftab, Peter Mollinga, and E. M. Shashidharan for useful resources and insights.

As Levinthal and March (1993: 110) note, "Conservative expectations, of course, will not always enhance the selling of learning procedures to strategic managers, but they may provide a constructive basis for a realistic evaluation and elaboration of the role of learning in organizational intelligence." Policy contexts are more complex than organizational ones. Given the appeal of incremental, rational, and deliberative change, it is tempting to overemphasize the importance of learning. In fact, learning is best viewed as a modest element in a politicized process of social change and control.

What Is Organizational Learning?

There is a considerable body of literature on organizational learning, drawn largely from the fields of organizational sociology and management.[1] Organizations can be seen as learning "by encoding inferences from history into routines that guide behavior" (Levitt and March 1988: 320) or by "improving actions through better knowledge and understanding" (Fiol and Lyles 1985: 803). According to these definitions, generating knowledge is not enough: learning also involves the use of knowledge of past experience to influence organizational practices.[2] Simply identifying shortfalls in organizational performance and assuming that the organization will use the information to improve performance is insufficient for effecting change.

In their widely cited work, Argyris and Schön (Argyris 1992; Argyris and Schön 1996) suggest that learning occurs at two basic levels in an organization—the single loop and the double loop. The single loop is "concerned primarily with effectiveness: how best to achieve existing goals and objectives, keeping organizational performance within the range specified by existing values and norms"; the double loop involves "inquiry through which organizational values and norms themselves are modified" (Argyris and Schön 1996: 22). Both single- and double-loop learning involve an iterative process in which information is processed in order to affect decisions.

Most models of organizational learning envision a cycle of four main steps, conceived in technocratic rather than political terms:

- Acquiring information about the organization and its environment
- Generating knowledge, by analyzing and interpreting information or reflecting on action
- Applying knowledge to organizational activity or experimenting with new ideas
- Encoding knowledge and experience into routines or memory.

Knowledge and action occur in tandem: knowledge can inform and guide action, and knowledge can be generated by reflecting on action. The cycle is iterative: in an ideal setting knowledge is constantly being modified based on new information and feedback, as a result of which, routines are constantly being refined.

Although learning is often viewed through such a rationalist and normative lens, in which the filtering and processing of stimuli are viewed as an objective and empirical process, learning processes are frequently subject to a series of social and institutional processes that are interpretative, symbolic, and power-laden. For example, policy and program evaluations can be undertaken for the symbolic purpose of legitimating existing activities rather than identifying areas for improvement. Organizations often engage in such ceremonial activities, some of which may involve the decoupling of information from decisions (Meyer and Rowan 1977; Feldman and March 1988). In addition, the information that receives attention in a decision process is not necessarily the information that would be most valuable for improving effectiveness or performance. As Cohen, March, and Olsen (1972) note in their "garbage can" model of decision making, it is sometimes the serendipitous confluence of actors and information in one place at one time that determines which decisions are made.[3]

Learning in a policy context is considerably more complex than learning in organizations for a variety of reasons, including, but not limited to the following:

- Policy making and implementation involve multiple organizational actors and political interests.
- Policy making occurs in a political arena, while implementation occurs in an administrative one. (This distinction is porous and sometimes artificial.)
- Time lags between policy making and policy implementation are long, often extending several years if not decades.
- Causal relations between policy choices and impacts are difficult to establish or predict, given the large number of possible variables and confounding factors.

What this suggests for examining learning in policy contexts is that it makes sense, at the very least, to examine policy processes over long time frames and to have conservative expectations about the potential for actual learning. It also suggests that a multistakeholder approach to examining policy can be useful for identifying options, alternatives, and differential impacts.

The seminal work of Pieter Glasbergen (1996, as described in Fiorino 2001: 324) on environmental policy in the Netherlands distinguishes three kinds of policy learning:

- Technical learning involves "a search for new policy instruments in the context of fixed policy objectives. Change occurs without fundamental discussion of objectives or basic strategies." This type of learning may be viewed as a form of single-loop learning, because it does not involve the questioning of basic goals, objectives, or problem definitions. The approach of policy makers to environmental problems is hierarchical and prescriptive, drawing on instruments with which they are familiar (regulations, oversight, and enforcement).
- Conceptual learning involves "a process of redefining policy goals and adjusting problem definitions and strategies." In this type of learning, policy objectives are debated and strategies reformulated. Because it requires new conceptualizations of problems and objectives (such as sustainability and ecological modernization), this type of learning may be viewed as a form of double-loop learning.

■ Social learning involves both technical and conceptual learning, but "it emphasizes relations among actors and the quality of the dialogue." It stresses the importance of multistakeholder views and information for improving both technical and conceptual learning.

These three types of learning can be viewed as complementary rather than mutually exclusive. They are linked by the fundamental precept of policy learning of iterative attention to and deliberation on policy issues. In technical learning, such iteration and deliberation are focused on improving policy instruments and their implementation; in conceptual learning, they emphasize policy goals and objectives. In social learning, the broader question of how multiple perspectives can enhance dialogue becomes more important.

Glasbergen's typology suggests that the institutional context of learning is crucial, because it shapes the relationships between actors and the quality of social learning. It is, therefore, impossible to separate learning about environmental policy from the legislative, legal, administrative, and democratic institutions that frame it. This linkage implies the need for reservation and caution in attempts to transfer policy successes in one country to another, given the variation in institutional contexts.

The "ideal-type" categorization by Glasbergen downplays at least two key factors that affect every stage of a learning cycle: bounded rationality and relationships of power. First, organizations and policy makers are limited in terms of the information they can collect and their ability to analyze and interpret it.[4] They are only "boundedly rational," unlikely to be able to make informed decisions without the participation of other stakeholders who possess other relevant information (March and Simon 1958). This is especially true of environmental policy contexts, in which there are multiple and interdependent issues concerning natural resources, public health, the economy, and environmental justice.

Second, relations of power frame the context in which policy making and implementation occur. In environmental policy contexts, conventional divisions between industry and environmental groups tend to portray policy choices as zero-sum games, pitting one set of technical experts against another. The "winners" in such cases are adept at advancing fixed policy positions (Nilsson 2005). There may be little social learning in such instances, however, given the absence of deliberative dialogue between actors. Understanding the social and political context of policy making, including relationships of patrimonialism and clientelism, is thus important for identifying possible avenues for and barriers to learning.

Given that each actor brings to the table a unique constellation of cognitive capacities and institutionalized relationships of power, learning is likely to be severely limited by these constraints of capacity and power. An approach to policy learning that involves collective engagement among actors has the potential to reduce technical uncertainty while providing openings for addressing political conflict. This approach resonates with the broader discussion in this volume on the necessity for an adaptive and inclusive methodology for strategic environmental assessment.

Framing Policy Learning

While it is possible to draw theoretical distinctions among the technical, conceptual, and social types of policy learning, actually observing policy learning in practice is exceedingly difficult. How can one gauge, for example, the extent to which dialogue and deliberation by diverse actors (that is, social learning) or simple political coercion accounted for a policy change? Does empirical evidence and technical information (that is, technical learning) actually influence policy makers, or do they tend to make policies based on convictions and political interests?

To determine whether a policy change was a result of learning, one would need empirical evidence to link behavioral change by policy makers or implementing agencies to information gathered and assessed as part of a deliberate or routinized process. One would need to identify who was learning, what was learned, and how it was learned while ruling out other plausible explanations for policy change.

These methodological demands cannot be met in a chapter such as this, which relies entirely on secondary data. The aim here is more modest: to provide a preliminary conceptual framing for policy learning, followed by an examination of two developing country cases based on limited secondary information.[5]

The framework draws on Glasbergen's premise that policy learning is inherently social, in that relationships and dialogue between actors are important even in technical discussions. Such a process relies on at least three themes that emerge repeatedly throughout this volume:

- *Agenda or priority setting.* If an environmental issue is deliberated by policy makers through a process based either on the systematic collection and consideration of empirical evidence (an adaptive management model) or an inclusive process of community participation (an inclusive management model), a policy learning process is at work.[6] In contrast, if the discussion of an environmental issue and its presence on the agenda is ad hoc or primarily a result of individual or interest group politics, opportunities for learning are less likely. It is thus necessary to understand the political context within which policy priorities are established. This context is likely to vary both across and within states. Key questions include the following: What are the mechanisms through which environmental priorities are identified and placed on policy agendas? Through which mechanisms do such priorities receive repeated attention that enables incremental improvement over time? Who has access and influence in setting these priorities? Adaptive or inclusive approaches do not necessarily precede agenda setting; in practice they are more likely to follow it. A key concern for learning purposes is thus how to focus repeated or iterative attention to a set of policy issues.

- *Stakeholder access and representation in policy formulation.* If policy learning is a deliberative and multistakeholder process, it is necessary to pay attention to how participation is operationalized and whether it is simply a consultative process (in which participants have limited influence) or a negotiated one (in which participants have greater influence). Key questions include the following: Through what mechanisms are multiple stakeholders brought into policy-making processes? Who participates in the process, and who has the power to choose or identify participants? Is their participation limited to "voice" or do they have actual "influence"?[7]
- *Accountability in implementation.* A key challenge for policy making in developing countries lies in implementation, which can be hindered by a range of factors, including a lack of resources and enforcement of laws and regulations. The presence of accountability mechanisms, such as transparency and right-to-know laws, monitoring and evaluation systems, inspectors general, and forms of judicial redress, can have an impact not only on implementation but also on the seriousness with which stakeholders engage the policy-making process itself. Policy learning requires feedback loops through which implementation results are linked to a new round of policy making. Key questions include the following: Once a set of policies is developed, what mechanisms ensure implementation? What forms of redress and conflict resolution are available to stakeholders? How are the effects of policies and their implementation monitored and fed back to policy makers and other stakeholders? Such scrutiny and iterative attention are important for both learning and accountability.

Attention to mechanisms in each of these three areas can arguably foster a climate that is more amenable to policy learning.[8] However, although agenda setting, representation, and accountability appear to be central to policy learning, they are not sufficient conditions for policy learning, given the political nature of policy making. They may, nonetheless, help overcome two key constraints on learning raised in the previous section: bounded rationality of policy makers and imbalances in relationships of power. These components of policy learning are depicted in figure 7.1. At the center of the figure is the conventional technical learning process, which is embedded in a much more complex conceptual and social learning environment that includes the three key elements described above.

Cases of Policy Learning

A key challenge facing developing countries face is the need to integrate economic development with environmental quality in a way that recognizes the trade-offs (Tobin 2003; Pearce 2004) while building supportive governance structures and the capacities of legislatures, bureaucracies, and courts for effective policy making (Durant, Fiorino, and O'Leary 2004). Many developing countries have good

FIGURE 7.1
Policy Learning as a Technical, Conceptual, and Social Process

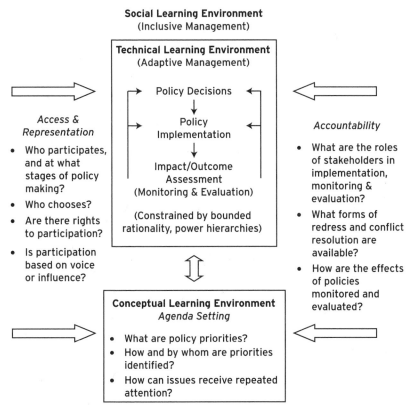

Social Learning Environment
(Inclusive Management)

Technical Learning Environment
(Adaptive Management)

Policy Decisions

Access & Representation
- Who participates, and at what stages of policy making?
- Who chooses?
- Are there rights to participation?
- Is participation based on voice or influence?

Policy Implementation
↓
Impact/Outcome Assessment
(Monitoring & Evaluation)

(Constrained by bounded rationality, power hierarchies)

Accountability
- What are the roles of stakeholders in implementation, monitoring & evaluation?
- What forms of redress and conflict resolution are available?
- How are the effects of policies monitored and evaluated?

Conceptual Learning Environment
Agenda Setting
- What are policy priorities?
- How and by whom are priorities identified?
- How can issues receive repeated attention?

Source: Author.

environmental legislation; the core challenges lie in implementation (Turner and Hulme 1997; Brinkerhoff and Gage 2002).

A caveat is necessary before proceeding to the case material. The cases outlined next are not assessments of environmental policy or policy learning. Because of the reliance on secondary materials and the difficulties in observing and operationalizing learning at the policy level, they are intended only to illuminate policy learning challenges and questions. The tentative nature of this task is amplified by the fact that the case materials rarely, if ever, explicitly cite "policy learning." The connections drawn to learning are imputed in order to identify key issues for further empirical investigation.

Brazil: Making Policy in a Patrimonial Context

Much of the discussion on policy making in Brazil is characterized by repeated reference to an entrenched feature of politics: patrimonialism.[9] Patrimonialism includes various forms of clientelism, rent-seeking, patronage, and co-optation, usually for purposes of personal gain rather than group or ideological interests.[10] One skeptical observer of Brazil's political context argues that "[t]here has been such a concentration of power in the hands of the State" that "civil society has had very little room to organize itself" and that legislative and electoral processes fall far short of representing the plurality of interests in Brazilian society (Guimarães 2002: 231–32). He suggests that Brazil's environmental protection agency has always played a marginal role in environmental policies, because of its emphasis on solving case-by-case problems and emergencies after they arise rather than preventing them and because of its dependence on political figures to champion specific causes.

Other observers, while agreeing on the relatively weak role of federal environmental policy making in Brazil, have suggested that Guimarães' account does not pay sufficient attention to Brazil's federated structure, in which considerable regulatory authority falls to individual states. A more-nuanced picture emerges in Ames and Keck's (1997–98) study of environmental policy making in the states of Mato Grosso, Paraná, Pernambuco, and Rondônia.[11] They confirm the political nature of policy making and its clientelist form at both the federal and state levels, observing that while federal and state assemblies often convene commissions to investigate charges of environmental abuse, they rarely exercise effective oversight.

At the same time, Ames and Keck point to the need to understand how local institutions mediate the policy process. They identify four factors that affected policy learning in the four states through their impact on clientelism, multistakeholder participation, and capacity:[12]

- *Competitiveness in state politics.* In Pernambuco "a long history of competing elites has produced a highly competitive and politically organized society, in which clientelism is held somewhat in check by a mass politics that requires substantive results to maintain popular support." In Paraná private business activities are more highly valued than politics, and clientelism is somewhat less pervasive. In contrast, in Mato Grosso extreme competitiveness has led to violence and a "Colombia-like accommodation among the state's older elites." In Rondônia highly competitive politics are based on alliances built through clientelism rather than political parties. In short, clientelist politics and competitiveness vary in substance and form across states. Competitiveness can hold clientelism in check, opening up spaces for policy learning through deliberation and dialogue of competing groups, but extreme competitiveness can exacerbate clientelism and even lead to violence. The impact of competitiveness on policy learning depends heavily on political context.
- *Bureaucratic capacity and stability.* Competent bureaucracies (such as that in Paraná) are associated with greater seriousness in budgeting and planning processes, including processes related to environmental policy making, technical

learning, and implementation. Where bureaucracies are more politicized, newly elected mayors and governors are inclined to purge them of holdovers from the previous administration (as occurred in Pernambuco), and there is greater likelihood of indifference to the law. Regardless of the stability of bureaucratic structures, the experiences of all four states show that state "environmental organs were often left out of the policy-making loop, consulted only after governors had already made policy choices" and that jurisdictional uncertainty about which agency should undertake which tasks was "rampant" across the states. Under conditions of weak bureaucratic capacity and instability, individual power and connections rather than an iterative process of learning become important for policy making.[13]

■ *Monitoring and information transparency.* Government agencies often withhold information from other agencies or state environmental councils or develop projects with little consultation with other agencies, legislatures, or affected populations. Policy learning depends on the gathering and analysis of information, its interpretation for shaping policy agendas, and its availability for public scrutiny. Ames and Keck point out that the "main processors of policy information in industrial societies—legislatures, courts, political parties, unions, and the media—are ill equipped to play this role in Brazil." Evidence from other Latin American countries points to the importance of monitoring and public scrutiny. Colombia's Cauca Valley Corporation attributes the success of its efforts to reduce water pollution to cooperating with industry (in a highly patrimonial climate) while deemphasizing traditional command and control efforts. But a crucial influence on industry behavior was citizen pressure and negative publicity in the press, which listed top polluters, coupled with demands by parent transnational companies for environmental audits (Sánchez-Triana and Ortolano 2001). For purposes of policy learning, this experience suggests that a combination of a monitoring regime and transparency can improve implementation while shedding light on whether a policy is working.

■ *Openings for civic participation.* Across all four states, the public was largely uninformed about environmental policy issues during the 1990s; mass media coverage of environmental issues was ad hoc. Civic intervention took place almost exclusively through nongovernmental organizations (NGOs), most of which were small, volunteer based, and unprofessionalized (there is evidence of improvements in their capabilities). One of the best-known examples of NGO activism is their representation of environmental interests and indigenous people affected by the Planafloro project in Rondônia, which resulted in their formal representation on state councils for the project and on an independent evaluation committee.[14] The representation of NGOs has been contentious in other states. In Paraná "developers quickly learned to form their own NGOs, adopt the word *environmental* in their titles, and accept official invitations to represent environmental interests on policy-making councils." An implication for policy learning is that NGOs have played important roles in enabling civic engagement, raising the level of dialogue needed for policy deliberation (provided they have adequate capacity). However, they are but one set of actors among many that have the potential to contribute to policy dialogue by increasing civic engagement. These factors point to the highly political nature of environmental policy making. From the perspective of policy learning, they suggest that deliberative discourse on policy options is likely to be weak where these factors are weak. In other words, promoting policy learning in environmental policy making is confounded by various political factors, including clientelist behavior and incentives, limited bureaucratic capacity and stability, inadequate monitoring and information transparency, constrained spaces for meaningful civic participation (not just voice but actual influence), public scrutiny, and enforcement of laws. The conditions for policy learning may thus be viewed as the effects of political action. Improving these conditions requires political exertion rather than simply better management and technocratic effort.

Oliveira (2002) provides an example of efforts to overcome some of these constraints in an analysis of environmentally protected areas (área de proteçáo ambiental [APA]) in the state of Bahia.[15] Although his study looks at policy implementation rather than formulation, it is instructive for its insights on bureaucratic capacity and political support. Oliveira identifies four major obstacles to policy implementation in developing countries: lack of political and public support, funding for protected areas, institutional capacity, and cooperation and coordination at the local level. Despite these obstacles, between 1990 and 1998 the number of APAs in Bahia grew from 2 to 27, expanding the area covered by a factor of 130.

In Oliveira's analysis, a central challenge in Bahia concerned the weak influence of environmental agencies relative to that of the development agencies. The state environmental agency "horizontally decentralized" the administration of APAs by distributing implementation responsibility among several agencies (rather than seeking to do it itself), thus helping integrate environmental policies with development policies. Oliveira argues that development agencies needed to establish protected areas in order to gain approval for their development projects from state and multilateral funders.

A second key feature of implementation was that funding for APAs was not determined a priori through an allocation to the state environmental agency. Had this been the case, it would likely have resulted in failure, because the state agency was chronically underresourced and did not have the political clout to ask for more funds. Instead, public agencies, especially the influential development agencies, could draw on multiple sources for implementing their policies.

A third factor was increased competition among public agencies for creating and controlling APAs. This competition arose from the fact that environmental projects tend to attract international funding, thereby increasing the political leverage of agencies engaged in those projects. While this competition initially created distrust among agencies, it eventually led to increased cooperation in order to create more APAs. An unexpected benefit of this cooperation was improvement in bureaucratic capacity across participating agencies, which occurred because coordination and sharing of expertise are necessary for gaining political leverage in requesting more state funding.

The Bahian experience offers a number of tentative insights into the three key themes of this chapter. First, the story of APAs is a political one, not a technocratic one. It demonstrates that environmental issues received greater attention in priority setting by linking to broader economic and social change agendas. Public agencies were motivated to prioritize protected areas as a way of increasing their political power and financial resources, within a broader economic agenda of tourism promotion that was firmly entrenched in clientelist politics. Arguably, the process hinged on the "commanding role" of a three-time Bahian governor in crafting "an unusually coherent policy process" that brought in regional development through a clientelist process that shored up his political power (personal correspondence, Margaret Keck, Johns Hopkins University, October 2005). This is not dissimilar to the South African experience in the early 1990s (described below), in which the education sector was more successful than the environmental sector in garnering attention and resources by "politicizing" education.

Second, while the Bahian experience does not speak directly to broad stakeholder representation, it does demonstrate the potential power of "horizontal decentralization" to secure the participation of multiple public agencies. Stakeholder participation through interagency coordination helps address the problem of public bureaucracies operating in isolation from one another, particularly given the weak political clout typical of environmental agencies. Global stakeholders also played a role, particularly the efforts of pro-conservation advocates to harness global concerns about deforestation in the Amazon.

Finally, this case demonstrates that an accountability system of checks and balances was necessary for implementation of the protected areas policy. Staff of the state environmental agency and the state environmental council (which included NGOs) were required to approve protected areas as well as development projects; they could block those that did not satisfy certain environmental guidelines. These councils had not just voice but direct influence or power of approval.

Taken together, the work of Guimarães, Ames and Keck, and Oliveira suggests that the central themes discussed in this chapter—priority setting, stakeholder representation, and accountability—do not stand apart from the politics of policy making but are shaped by it. There is a dialectic relationship among these three components, their political context, and opportunities for policy learning.

South Africa: Making Environmental and Educational Concerns Relevant to Citizens

The case of South Africa is instructive in a study of policy learning because of the country's unprecedented policy development just before and during its first decade of democracy. South Africa has undergone a deliberate and ongoing shift from the centralized, technocratic, and discriminatory policy making of the apartheid era toward a more inclusive and equitable set of policies and policy processes. While the country boasts a highly progressive and inclusive environmental policy framework today, policy makers paid very little attention to environmental issues in the transitional years preceding democracy in 1994 (Orkin, Tshandu, and Dugard 1995).

Educational policy in the late 1980s and early 1990s followed a radically different trajectory.[16] The education sector experienced a range of policy initiatives that not only brought various policy stakeholders—universities, parastatal research centers, NGOs, industry, and political parties—together but also generated extensive public interest and debate, leading to a number of major reforms in postapartheid education.

Orkin, Tshandu, and Dugard (1995) offer a number of possible reasons for this difference, some of which are relevant to policy learning:

- *Politicization of an issue.* There was keen public interest in education, the relevance of which was apparent to all segments of society. The issue of education was highly politicized, in that it served as a key political platform for state actors and the democratic movement. In contrast, environmental actors focused on conservation and preservation of wilderness areas, which were of greater interest to white elites than to the majority of South Africans. They failed to politicize their issues broadly and thus to make them relevant to the larger public (by linking them to drinking water, health, and conflict, for example).

- *Proactive state involvement.* The state's involvement in education was proactive and mass based. The white-led Department of National Education conducted two major policy formulation initiatives. The black-led National Education Coordinating Committee convened a massive two-year National Education Policy Initiative. In contrast, the approach in environmental policy was driven either by narrow industrial interests or by the elitist conservation orientation of large NGOs such as the Endangered Wildlife Trust, the Southern African Nature Foundation, and the Wildlife Society of Southern Africa.

- *Research capacity.* Research mechanisms to inform policy formulation were established in education but not in the environmental arena. The National Education Coordinating Committee's policy efforts were supported through research at university-based Educational Policy Units; the African National Congress'

Education Department drew on recommendations from the Center for Educational Policy Development (CEPD), an NGO; the state commissioned research through the Human Sciences Research Council; and the business sector had its own education research centers. The CEPD is particularly interesting as a policy learning catalyst, because it worked "collaboratively with the community of policy analysts in universities, mass-based movements, and NGOs, especially those associated with the democratic movement" (Orkin, Tshandu, and Dugard 1995: 480).

- *High-level multistakeholder policy forums.* Multiple policy actors in the education sector were brought together for vigorous high-level policy debates. Through consultation with several hundred people from different constituencies, the CEPD developed a draft policy framework for the African National Congress (ANC). The National Training Board (NTB), a statutory body focused on national labor issues, also opened up its policy formulation process in response to pressure from labor groups, eventually setting up eight working groups to formulate NTB policy on education and training. Both the CEPD and the NTB exercises marked a shift from unilateral policy formulation by separate institutions characterized by standoffs among policy actors into relationships of exchange, engagement, and cooperation.[17] Even in these multistakeholder arenas, however, policy formulation remained dominated by elites.

Since the study by Orkin, Tshandu, and Dugard, there have been dramatic changes in environmental policy in South Africa, particularly the inclusion of environmental rights and justice in the constitution, the creation of a Consultative National Environmental Policy Process (CONNEPP), and the promulgation of the 1998 National Environmental Management Act. These successes in environmental policy making in the first decade after democratization appear, at least on preliminary examination, to bear some of the hallmarks of the factors identified above: politicization of environmental concerns into issues of rights and justice, proactive state involvement and forums for stakeholder engagement (through CONNEPP), and an increase in the research capacities on environmental issues of various actors, including government agencies, universities, and NGOs.

Such momentum is difficult to implement and sustain. A review of the consultative process under CONNEPP carried out by researchers at South Africa's Council for Scientific and Industrial Research and the City of Cape Town suggests that many of the challenges apparent in 1995 were still relevant to environmental policy making in 2004 (Rossouw and Wiseman 2004). These challenge include insufficient consultation with local government and local councils; failure to maintain stakeholder engagement and networks after completing the policy-formulation process; lack of ongoing consultation with civil society in the implementation and monitoring of policy; failure to follow up on key operational commitments in the policy; unfunded mandates for sustainability planning;[18] inadequate integration of environmental priorities with national priorities of poverty eradication and social transformation; and lack of mechanisms for policy learning and continuous improvement, including policy assessment and monitoring, as part of the policy process. Even less-critical observers point to the difficulties in implementing policy and enabling adaptive learning, noting that while South African water management policy "is considered internationally to be progressive, forward-thinking and ambitious," there is a strong need for a "strategic, adaptive approach to policy implementation" through a process that "allows for learning to be gathered along the way and fed back into improving the process" (MacKay, Rogers, and Roux 2003: 353–54).

These issues represent challenges for both adaptive and inclusive management. The list could apply to the environmental policy and implementation processes of most countries, developing or developed. As such, the experiences of environmental policy making in South Africa illuminate three lessons for the three themes of this chapter—the need to politicize environmental issues in order to get them onto policy agendas, the importance of building stakeholder capacity in order to have quality stakeholder representation, and the challenges of creating accountability systems that are not just about transparency of information but also about monitoring and follow-through.

Conclusions

The cases examined in this chapter identify a number of factors that are significant in constraining and enabling policy learning. The lessons learned should be taken as propositional and thus subject to empirical validation.

Priority or Agenda Setting

Policy learning, in the sense of iterative attention of policy makers to an issue, requires getting environmental issues on their agendas in the first place. This can be done in several ways:

- "Politicize" environmental issues, by linking them to broader issues of economic and social development and poverty eradication, in order to make it more likely that policy makers will take the risks of engaging them. Issues presented in highly technocratic terms of narrow interest (such as environmental preservation and conservation) are unlikely to nurture mass support if they do not appeal to the daily struggles of electorates. Effective politicizing was apparent in the case of education policy in the lead up to democracy in South Africa. While environmental policy was initially mired in preservation terms, it later underwent a significant shift to focus on issues of environmental justice, livelihood, and equity.
- Integrate the agendas of environmental ministries with those of more-influential ministries. A sustainable development agenda requires explicit links to issues such as food and income security, water scarcity, and conflict over resources.

The experience of protected areas in Bahia shows that integration between environment and economic development is possible even in a patrimonial and competitive policy context.

■ Create policy advocacy networks in order to get diverse perspectives on the agendas of policy makers not just once but repeatedly. Sustained NGO activism was central to the representation of environmental and indigenous organizations on a deliberative council in Brazil's Planafloro project, as well as to the critique of its predecessor project. In the case of protected areas in Bahia, NGOs and other actors who were members of state environmental councils had both administrative influence with respect to approval for projects and access to policy circles.

Stakeholder Representation in Policy Formulation

If policy learning is a deliberative and multistakeholder process of engagement on policies and ideas, it is necessary to pay attention to how participation is operationalized. Consultative process must give participants not just voice but influence. This is a problem of elite control of policy formulation and implementation (for example, through clientelism, rent-seeking behavior, and expert domination).

Several lessons can be drawn from the case studies:

■ The capacity for policy research and analysis across all types of policy actors (governments, civil society organizations, industry, universities) is crucial to policy learning. The impact of such capacity is most apparent in the case of South Africa's education sector, where the major policy actors had access to research units and were able to participate in national forums that directly affected policy. This stood in stark contrast to environmental policy processes, in which there was limited research capacity, except perhaps in industry (this has changed significantly in recent years).

■ Public forums for policy debate vary in significance for policy learning. Forums that are consultative in nature and in which participants have formal voice but cannot directly shape policy are sometimes exercises in public relations rather than policy learning. In the South African education sector, the forums established through the National Training Board and the CEPD were premised on the notion that managed conflict and tension are necessary for democratic policy making. They marked a shift from a process of unilateral policy formulation by separate institutions into relationships of exchange and cooperation.[19] Similarly, the South African Consultative National Environmental Policy Process made considerable headway in creating political space for policy debate. It has been critiqued, however, as having been undermined by insufficient follow-through in maintaining stakeholder engagement networks and in integrating with national priorities on poverty eradication.

■ The participation of multiple public agencies on environmental policy issues is necessary for reconciling interagency tensions while improving coordination and resources. The Bahian experience with "horizontal decentralization" improved interagency coordination. Although a key incentive was access to funding and policy influence at the state level, such coordination increased bureaucratic capacity and competence. Differences in bureaucratic competence can vary significantly within a single country, as the four-state study of Brazil reveals. Overcoming the constraints of bounded rationality in public decision making requires improving interagency coordination.

Accountability in Implementation

Policy learning requires feedback loops through which implementation is linked to policy making and through which constituents can hold policy makers to account for both follow-through and the consequences of policy decisions. This is the least-developed dimension of the case studies.

The case studies—and the literature on accountability—nevertheless offer some insights:

- Effective transparency mechanisms make information available to citizens in ways that allow it to influence their political choices.[20] They provide complete information about activities and options before key decisions are made, in local languages, culturally appropriate formats, and ways that are readily accessible and affordable (Nelson 2001). In this sense transparency is necessary not only for enhancing policy dialogue but also for monitoring the activities of public agencies. Examples include public right-to-know legislation and citizen oversight of public budgeting activities.

- Monitoring and evaluation of the policy-making and implementation processes is costly and rare but important for policy learning. The experiences with state environmental councils in Bahia and the Planafloro deliberative council in Brazil suggest that citizen groups can play central roles in monitoring the impacts of policies and monitoring how and if they are implemented. Indeed, part of the critique of consultative processes in environmental policy making in South Africa has centered on insufficient procedures and funding for monitoring and evaluation, by both public agencies and citizen groups.

- Supporting media scrutiny of policy and implementation issues can enhance public monitoring while strengthening accountability. There is evidence that the public and industry respond to sustained and pointed media scrutiny. Publication by the local newspaper of the names of top polluters in Colombia's Cauca Valley, combined with citizen pressure and policies of parent transnational companies that called for environmental audits, motivated better pollution control efforts (Sánchez-Triana and Ortolano 2001). Legislative reporting requirements can enhance such scrutiny (as noted above).

- Maintaining stakeholder engagement and networks after completing a policy formulation exercise is important for long-term learning and dialogue. The experiences with educational and environmental reform in South Africa suggest that one-off policy engagements are unlikely to lead to sustained policy improvement.

In closing, four more general lessons arise repeatedly regarding policy learning. First, policy learning may be enhanced where agendas and priorities are shaped by multiple stakeholder perspectives, including those of public agencies, ministries, and civil society groups. The politicization of an issue in broad developmental and social terms (rather than only in environmental terms) seems crucial for repeated policy attention. Second, given the iterative nature of policy formation and the limitations of bounded rationality, policy learning may benefit from participation of stakeholders at all stages of a policy process, from agenda setting through implementation, monitoring, and evaluation.

Third, accountability in policy learning and implementation relies not only on monitoring and evaluation but also on effective transparency and enforcement mechanisms. Transparency refers not only to the provision of information before key decisions are made but also to the presentation of that information in forms users can understand. This combination appears to be an important ingredient for sustained and informed policy debate and thus for social learning. It is especially important in countries where right-to-know legislation and legal forms of redress are weak. In the long term, efforts to strengthen the rule of law may also enhance policy learning, by making policy makers and agency administrators more responsive to citizen concerns.

Finally, the factors or conditions identified above (on agenda setting, stakeholder access and representation, and accountability) suggest that there are no simple institutional mechanisms for ensuring learning. Moreover, these conditions are not sufficient for policy learning, largely because policy formulation and implementation are highly political processes. Nonetheless, the cases do provide some preliminary evidence that the 10 conditions cited here can ease two key constraints on policy learning: bounded rationality and hierarchies of power. Enhancing the negotiated and deliberative dimension of policy making—by involving multiple actors in roles of influence, supported by capacities for research and monitoring, in ongoing and iterative dialogue—can help deal with both problems. The cases also suggest that opportunities for policy learning can be enhanced even in circumstances of adversarial power relations and clientelism. Expectations should be moderated, however, by the fact that policy learning is but one component in broader processes of social and political transformation.

Notes

1   The terms *organizational learning* and *learning organization* are often used interchangeably in the literature. *Organizational learning* is more widespread in the analytical literature; *learning organization* is more normative (Denton 1998; Senge 1990).

2 Learning is only one of many processes yielding organizational change. For example, changes in organizational routines can also be brought about through new laws that mandate procedural changes. Moreover, learning is not always an intentional process, and it does not always lead to improvements in an organization's performance (Levitt and March 1988; Scott 1992).

3 The garbage can model is discussed in chapter 3 of this volume.

4 Limitations in cognitive capacity can lead to "superstitious" or "ambiguous" learning. Superstitious learning arises when individuals incorrectly deduce that a specific action led to a particular outcome (March and Olsen 1988). Ambiguous learning occurs when an outcome is so poorly understood that multiple explanations emerge. The learning that results is ambiguous, because meaning is imputed even though it may not be clear what exactly happened, why it happened, or whether what happened was good or bad (March and Olsen 1988). Such ambiguity is more likely in policy contexts than in organizational ones.

5 Identifying actual instances of learning is also difficult in developed country contexts. In their study of accountability and performance in public bureaucracies in the United States, for example, Gormley and Balla (2004: 18) observe that "with few exceptions, legislators have ignored performance data, even from agencies that have produced relatively useful and complete information." Efforts by legislators to advance fixed policy positions, rather than those amenable to change through learning, are also common in Sweden (Nilsson 2005).

6 Chapter 3 examines both models.

7 The distinction between voice and influence is significant. In a widely cited background paper prepared for the *Human Development Report 2002*, Goetz and Jenkins (2001: 9) note that "simply listening to these voices [of the poor], and doing nothing to respond to their insistent demands has discredited the idea that promoting voice is central to improving human well-being." The World Bank's *World Development Report* (2003b: 79) extends this reasoning to accountability, noting that "voice is not sufficient for accountability; it may lead to answerability but it does not necessarily lead to enforceability."

8 This assumption is contestable and potentially tautological. Broad stakeholder participation does not guarantee learning, particularly where it is used primarily for public relations. Similarly, heavy-handed accountability requirements can restrict learning by limiting risk taking and innovation. The conclusions drawn from these assumptions should thus be treated as propositional.

9 The author is indebted to Professor Margaret Keck for detailed comments on this section of the chapter.

10 The challenges of patrimonialism and regulatory capture are, of course, not limited to Brazil. An institutional assessment of Colombia's environmental management system, for example, finds that despite the creation of numerous mechanisms for public participation in environmental policy formulation and implementation, regulatory capture by private sector interests has been significant (Blackman and others 2004).

11 The discussion of these states is helpful for identifying key political factors that shape opportunities for policy learning. Given both the limited and dated nature of the available secondary material, it should not be extended beyond this purpose.

12 Except where otherwise noted, this section relies on and quotes from Ames and Keck (1997–98: 25–31).

13 Even in the most successful state for environmental policy (Paraná), policy making "remains highly dependent on individuals: the installation of the governor's brother as head of the state's new environmental organ halted new initiatives" (Ames and Keck 1997–98: 28).

14 Critics contend that while this inclusion gave some NGOs a formal voice in the process, it provided little actual influence, because the council was not directly involved in planning, it failed to include a number of large and well-organized interest groups, and it did not provide NGOs with the capacity building needed to participate effectively. This is an example of the distinction between voice and influence raised earlier in the chapter. Later stages of the project appear to have partially responded to these concerns by increasing NGO participation in supervision, monitoring, and evaluation while increasing support for indigenous communities. For a self-critical review of the Planafloro project, including how NGOs were involved, see World Bank (2003a).

15 Brazil's National Environmental Council defines an APA as an area "destined to preserve the environmental quality and natural resources in a certain region in order to improve the quality of life of the local population and to protect regional ecosystems" (cited in Oliveira 2002: 1732).

16 Unless otherwise noted, this comparison of environmental and educational policy formulation draws on Orkin, Tshandu, and Dugard (1995).

17 This national consultative approach was later adopted in the environmental arena.

18 Some large municipalities, such as the City of Cape Town, have been better able to establish stakeholder consultation processes for formulating and implementing municipal environmental policies. Rossouw and Wiseman (2004: 138) call this a "paradigm shift" from "denying access to information . . . to the current understanding that information can provide a means to empower citizens."

19 Such forums should not be idealized. Miraftab's (2003) work on housing policy formulation and implementation in South Africa shows that even under seemingly positive conditions, the practice of participation can easily degenerate into a zero-sum game, particularly in the absence of efforts to strengthen the capacities of weaker actors to shape agendas and claim their rights.

20 For a broader discussion of transparency, see Nelson (2001), Fung and others (2004), and Herz and Ebrahim (2005).

References

Ames, Barry, and Margaret E. Keck. 1997–98. "The Politics of Sustainable Development: Environmental Policy Making in Four Brazilian States." *Journal of Interamerican Studies and World Affairs* 39 (4): 1–40.

Argyris, Chris. 1992. *On Organizational Learning.* Cambridge, MA: Blackwell.

Argyris, Chris, and Donald A. Schön. 1996. *Organizational Learning II: Theory, Method, and Practice.* Reading, MA: Addison-Wesley.

Blackman, Allen, Sandra Hoffman, Richard Morgenstern, and Elizabeth Topping. 2004. *Assessment of Colombia's National Environmental System (SINA).* Washington, DC: Resources for the Future.

Brinkerhoff, Derick W., and James D. Gage. 2002. "Natural Resources Management Policy in Africa: Implementation Challenges for Public Managers." In *Environmental Policy and Developing Nations*, ed. S. S. Nagel, 74–114. Jefferson, NC: McFarland & Company.

Cohen, Michael D., James G. March, and Johan P. Olsen. 1972. "A Garbage Can Model of Organizational Choice." *Administrative Science Quarterly* 17 (1): 1–25.

Denton, John. 1998. *Organizational Learning and Effectiveness.* London: Routledge.

Durant, Robert F., Daniel J. Fiorino, and Rosemary O'Leary. 2004. "Environmental Governance Reconsidered: Challenges, Choices, and Opportunities." Cambridge, MA: MIT Press.

Feldman, Martha S., and James G. March. 1988. "Information in Organizations as Signal and Symbol." In *Decisions and Organizations*, ed. J. G. March, 410–28. Cambridge, MA: Basil Blackwell.

Fiol, C. M., and M.A. Lyles. 1985. "Organizational Learning." *Academy of Management Review* 10 (4): 803–13.

Fiorino, Daniel J. 2001. "Environmental Policy as Learning: A New View of an Old Landscape." *Public Administration Review* 61 (3): 322–34.

Fung, Archon, David Weil, Mary Graham, and Elena Fagotto. 2004. "The Political Economy of Transparency: What Makes Disclosure Policies Effective?" Ash Institute for Democratic Governance and Innovation, John F. Kennedy School of Government, Harvard University, Cambridge, MA.

Glasbergen, Pieter. 1996. "Learning to Manage the Environment." In *Democracy and the Environment: Problems and Prospects*, eds. W. M. Lafferty and J. Meadowcroft, 175–93. Cheltenham, UK: Edward Elgar.

Goetz, Anne Marie, and Rob Jenkins. 2001. "Voice, Accountability and Human Development: The Emergence of a New Agenda." Background paper for the *Human Development Report 2002*. New York: United Nations Development Programme.

Gormley, William T., and Steven Balla. 2004. *Bureaucracy and Democracy: Accountability and Performance*. Washington, DC: Congressional Quarterly Press.

Guimarães, Roberto P. 2002. "The Bureaucratic Politics of Environmental Policy Formation in Brazil." In *Environmental Policy and Developing Nations*, ed. S. S. Nagel, 227–44. Jefferson, NC: McFarland & Company.

Herz, Steve, and Alnoor Ebrahim. 2005. "A Call for Participatory Decision Making: Discussion Paper on World Bank–Civil Society Engagement." Report commissioned by the civil society members of the World Bank–Civil Society Join Facilitation Committee, World Bank, Washington, DC. http://siteresources.worldbank.org/CSO/Resources/World_Bank_Civil_Society_Discussion_Paper_FINAL_VERSION.pdf.

Levinthal, Daniel A., and James G. March. 1993. "The Myopia of Learning." *Strategic Management Journal* 14 (Winter): 95–112.

Levitt, Barbara, and James G. March. 1988. "Organizational Learning." *Annual Review of Sociology* 14: 319–40.

MacKay, H. M., K. H. Rogers, and D. J. Roux. 2003. "Implementing African Water Policy: Holding the Vision While Exploring an Uncharted Mountain." *Water SA* 29 (4): 353–58. http://www.wrc.org.za.

March, James G., and Johan P. Olsen. 1988. "The Uncertainty of the Past: Organizational Learning under Ambiguity." In *Decisions and Organizations*, ed. J. G. March, 335–58. Cambridge, MA: Basil Blackwell.

March, James G., and Herbert A. Simon. 1958. *Organizations*. New York: John Wiley.

Meyer, John W., and Brian Rowan. 1977. "Institutionalized Organizations: Formal Structure as Myth and Ceremony." *American Journal of Sociology* 83 (2): 340–63.

Miraftab, Faranak. 2003. "The Perils of Participatory Discourse: Housing Policy in Postapartheid South Africa." *Journal of Planning Education and Research* 22 (3): 226–39.

Nelson, Paul J. 2001. "Transparency Mechanisms at the Multilateral Development Banks." *World Development* 29 (11): 1835–47.

Nilsson, Mans. 2005. "Learning, Frames and Environmental Policy Integration: The Case of Swedish Energy Policy." *Environment and Planning C, Government and Policy* 23 (2): 207– 26

Oliveira, Jose Antonio Puppim de. 2002. "Implementing Environmental Policies in Developing Countries through Decentralization: The Case of Protected Areas in Bahia, Brazil." *World Development* 30 (10): 1713–36.

Orkin, Mark, Zwelakhe Tshandu, and Jackie Dugard. 1995. "Policy Research and Formulation: Two Case Studies." In *Managing Sustainable Development in South Africa*, eds. P. FitzGerald, A. McLennan, and B. Munslow, 469–507. Cape Town: Oxford University Press.

Pearce, David. 2004. "Growth and the Environment: Can We Have Both?" In *Environment Matters*, 14–15. Washington DC: World Bank.

Rossouw, Nigel, and Keith Wiseman. 2004. "Learning from the Implementation of Environmental Public Policy Instruments after the First Ten Years of Democracy in South Africa." *Impact Assessment and Project Appraisal* 22 (2): 131–40.

Sánchez-Triana, Ernesto, and Leonard Ortolano. 2001. "Influence of Organizational Learning on Water Pollution Control in Colombia's Cauca Valley." *Water Resources Development* 21 (3): 493–508.

Scott, W. Richard. 1992. *Organizations: Rational, Natural, and Open Systems*. Englewood Cliffs, NJ: Prentice Hall.

Senge, Peter M. 1990. *The Fifth Discipline: The Art and Practice of the Learning Organization*. New York: Currency Doubleday.

Tobin, Richard J. 2003. "Environment, Population, and the Developing World." In *Environmental Policy: New Directions for the Twenty-First Century*, eds. N. J. Vig and M. E. Kraft, 347–69. Washington, DC: Congressional Quarterly Press.

Turner, Mark, and David Hulme. 1997. *Governance, Administration and Development: Making the State Work*. West Hartford, CT: Kumarian.

World Bank. 2003a. "Implementation Completion Report (CPL–34440) on a Loan in the Amount of US$167.0 Million to the Federative Republic of Brazil for a Rondonia Natural Resources Management Project (Loan 3444–BR)." Washington, DC.

———. 2003b *World Development Report 2004: Making Services Work for Poor People*. Washington, DC: World Bank.

C H A P T E R 8

Using Strategic Environmental Assessments to Design and Implement Public Policy

Kulsum Ahmed and Ernesto Sánchez-Triana

HOW CAN A STRATEGIC ENVIRONMENTAL ASSESSMENT (SEA) shape the design and implementation of public policy in the direction of greater sustainability? Under what circumstances can SEA strengthen environmental governance? Two very different examples—one relating to the functioning of regional development corporations, the other to action on the air pollution agenda, both documented in the Colombia Country Environmental Analysis (CEA) (Sánchez-Triana, Ahmed, and Awe 2007)—show the potential of SEA processes to shape the design and implementation of public policies aimed at alleviating poverty, promoting economic growth, and preventing environmental degradation.[1]

One important area identified in the Colombia CEA as needing attention to improve the functioning of regional development corporations (*corporaciónes autónomas regionales* [CARs]) was the importance of aligning national environmental priorities and regional environmental expenditures (World Bank 2006; Sánchez-Triana, Ahmed, and Awe 2007).[2] According to these needs, the government designed a system of performance indicators for CARs, adopted as Presidential

Kulsum Ahmed is lead environmental specialist and team leader of the Environmental Institutions and Governance program at the World Bank. Ernesto Sánchez-Triana is a senior environmental specialist at the World Bank. The authors would like to thank Shahid Yusuf, Leonard Ortolano, and Fernando Loayza for their insightful comments on earlier versions of this chapter.

Decree 1200 of 2004. The decree requires the CARs to measure the impact of their environmental investments, through 10-year regional environmental management plans and 3-year action plans (*Planes de Acción Trienal* [PATs], thus promoting greater accountability from CAR directors. From a list of environmental quality impact indicators of national importance, CAR directors are asked to select those that apply to their region.[3] This procedure strengthens the linkage between national priorities and regional implementation while allowing the regions to choose their own priority areas. At the national level, the Ministry of the Environment reviews progress on PAT goals. The decree includes a provision for removing the CAR director if the goals set forth in the PAT are not met.

The second example relates to the air pollution agenda. The Colombia CEA highlighted indoor and outdoor air pollution as key priorities (World Bank 2006; Sánchez-Triana, Ahmed, and Awe 2007), but the government was not inclined to immediately revise the national air quality standards developed in the 1980s. Wide media coverage of the results of the Colombia CEA, however, resulted in a broad public debate, which was taken up by politicians during the election campaign. One senator placed white blankets over Bogotá and widely publicized the change in the color of the blankets, comparing it with the effect on people's lungs and quality of life. Open public debate increased the number of champions for revising air quality standards (Chavarro Vásquez 2007). In 2007, the first air pollution control bill was discussed in the national congress (interview with Senator Nancy Patricia Gutierrez, President of the National Congress of Colombia, August 15, 2007).

Both these cases and related dialogue with the government illustrate the importance of internal champions and constituencies in moving reforms forward. They also illustrate that policy design and implementation lie on a continuum and that policies can be influenced during a current period of policy reform or on a longer time scale in a future period of policy reform. In the first case, the analysis in the Colombia CEA was already available when an opportunity to effect reform came along. This opportunity was created by multiple factors, including the importance President Alvaro Uribe placed during his election campaign on improving the efficiency of the CARs in delivering services (Uribe 2002a) and alleged cases of corruption among CARs (Uribe 2002b). The information in the CEA was timely and hence used in the design of the reform. In the second case, the wide dissemination of the results of the Colombia CEA and the ensuing media and political take-up and dialogue with civil society helped create constituencies for change at a crucial time (elections). The first case is an example of seizing an opportunity by being in the right place at the right time and being prepared; the second case is an example of creating an opportunity to effect reform by generating information and providing a forum for debate in order to create constituencies.

This chapter draws on the authors' experiences, including the two case studies described above, as well as lessons learned from other chapters in this volume to propose a new conceptual and methodological framework for applying SEA to policies. The chapter proposes a useful tool—namely, institution-centered SEA—for designing and implementing equitable and environmentally sustainable policies and strengthening environmental institutions and governance. In addition to applying analytical and participatory approaches typical within SEA frameworks (OECD 2006), institution-centered SEA incorporates a third pillar designed to enhance learning and continuous improvement of policy design and implementation (figure 8.1). The World Bank is currently piloting this institution-centered SEA in different sectors and regions of the world with results from the pilots being used to refine the approach (World Bank 2005a).[4]

As suggested by the examples above, institution-centered SEAs yield recommendations with both short- and long-term effects. Recommendations with short-term effects help identify current environmental priorities and the technical and institutional solutions that can address them; they enable actors to seize opportunities to influence already-initiated policy reform. Recommendations with longer-term impact help strengthen institutions, so that opportunities are continuously created to raise issues that matter to the weakest or most vulnerable citizens, thereby promoting social accountability of public officials. Raising issues that matter to the weaker members of society helps create opportunities for future reform, so that there is continuous improvement of public policies in the direction of greater sustainability. Because governance is the act or manner of governing, this suggests that the approach to SEA for policies presented in this chapter could be used as an instrument for sound and sustainable governance.

This chapter is divided into five sections. The first section briefly describes the application of strategic environmental assessment for policies and the limited influence of previous approaches in effecting change in the context of policy reform. The second section presents some important concepts with respect to policy formation and their implications on SEA for policies. The third section describes the conceptual framework of the proposed SEA approach and its components. The fourth section provides good-practice applications

FIGURE 8.1
The Three Pillars of Institution-Centered SEA

Analytical work	Participatory approaches	Learning processes

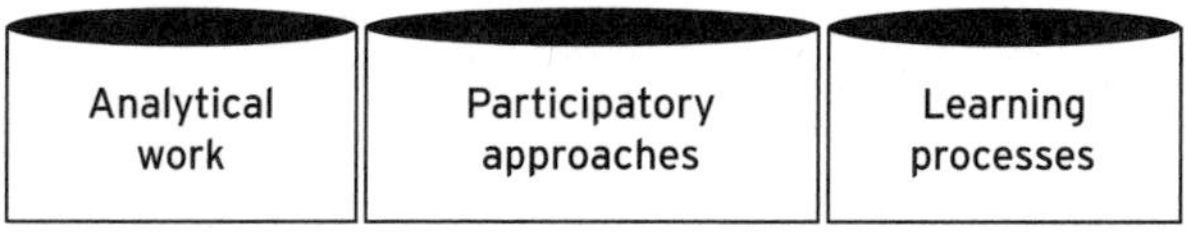

Source: Authors.

of various components of the proposed approach. The last section summarizes the chapter's main conclusions.

SEA for Policies

SEA was initially based on the application of environmental impact assessment methodologies to the environmental impact of groups of investment projects clustered in programs and of land-use zoning and regional plans. Considerable experience exists in applying SEA to investment programs and land-use zoning plans (Sadler and Verheem 1996; Fischer 2002; Noble 2002; Partidario 2002; Ahmed, Mercier, and Verheem 2005). The international review by Dalal-Clayton and Sadler (2005) shows the wealth of ongoing SEA activity. Donors are also trying to harmonize approaches to SEA and identify new opportunities for its application (OECD 2006).

Chapter 1 discusses the traditional methodologies applied in SEAs for plans and programs and provides some examples of their application. Typically, the initial step involves identifying potential environmental impacts. An assessment then determines which impacts are most significant. Mitigative actions are then proposed, along with monitoring frameworks. Public consultation typically takes place to help identify potential impacts.

There are fewer examples of the application of SEA to policies. As Ortolano notes in chapter 2, the influence of policy-level SEA on policy formulation and implementation depends heavily on process integration issues, especially those concerned with when SEAs begin relative to the policy-formulation and implementation process and how often SEA teams and policy designers interact. According to Ortolano, national assessments of the application of SEA in Canada and the Netherlands indicate that agencies often prepare these assessments on a pro forma basis. Assessments often occurred late in the policy-formulation process, and agencies were allowed to marginalize environmental assessment requirements without penalty. Key factors affecting policy SEA in Canada and the Netherlands appear to be linked to difficulties in applying impact assessment methodologies to policy proposals, the absence of cross-sectoral agencies with responsibility for overseeing compliance with SEA requirements, and lack of commitment to SEA on the part of top-level agency officials.[5]

Traditional SEA methodologies, developed for plans and programs, are difficult to apply to policy formation. The first step (identifying environmental impacts) is difficult because policies do not have defined boundaries and scenarios developed are subject to path dependence (the influence of historical events on the present and the future). Public consultations are also problematic, because equal representation of those most affected by environmental degradation is difficult to ensure given vested interests and elite capture. Traditional SEA also represents an assessment at a point in time rather than a continuous process, with SEA agents often focusing more on the preparation of a report than on how to influence decision makers. Because policy formation is a continuous process, with windows of opportunity to effect change, SEA methodologies for policies need to focus on taking advantage of opportunities as they occur.

Policy Formation and Challenges for Applying SEA

Policies differ from investment programs or land-use plans in several ways. Each of these differences is described below.

Policy Formation Is Continuous

Feldman and Khademian (chapter 3) show that given the nonlinear process of policy formation, SEA could be applied as a tool with which to take advantage of windows of opportunity that occur when there is a concurrence of issues, problems, solutions, and people in the context of policy decision making. They advocate that SEA be designed to seize opportunities, as was done in Colombia to improve the CARs. They present two policy-formation models, the adaptive management model, which strives to learn from previous experience and support incremental improvements in policies, and the inclusive management model, which brings different stakeholders' viewpoints together continuously throughout the policy process. Their analysis suggests that an SEA approach for policies should focus on promoting continuous improvements in policy design and a learning culture based on the involvement of many stakeholders.

Historical Events Shape and Lock in Policies

The influence of historical events on the present and the future (path dependence) plays a key role in shaping and locking in public policies.[6] However, as development contexts change, so may the environmental issues that affect stakeholders. The air pollution example from Colombia illustrates this point. As a result of Colombia's shift from a primarily rural economy to a highly urbanized country during the course of four decades, urban air pollution is now a priority. The mismatch between resource allocation and priorities is understandable given historical factors. Path dependence suggests that any attempt to shape institutional change requires periodic reevaluation of the goal itself rather than an incremental approach to measuring results that assumes the goal is still valid. Analysis is important, as is directly asking affected stakeholders what is working and what is not. Most important, however, is a commitment to be open, to learn from the analysis and the answers, and to act on this information.[7] This suggests that rather than assume that the general perception of current priorities is still valid, an SEA approach for policies needs to reevaluate environmental priorities based on both participatory methods as well as on an objective analytical basis.

Elite Capture and Vested Interests Influence Policy Design and Implementation

Elites are small groups with partial or total control over policy and the economy. They include large landowners and business owners, bureaucrats, politicians in authoritarian regimes, unions, religious leaders, and ruling castes, among others. They create formal rules that maximize their payoff by affecting policies and generating or blocking institutional reforms (Easterly 2001).[8] Elites are not necessarily monolithic; they include different groups with partly divergent interests. Hence, departing from traditional SEA methodologies to include political economy analysis is important in institution-centered SEAs.[9]

Bardhan (1989, 2004) and others have discussed the tenacity of vested interests and the difficulties in bringing about institutional change. Collective action by the nonelite (possibly supported by dissident groups of the elite), international pressure, or exogenous shocks can sometimes make reform feasible even in countries with weak institutions, as it has in the case of national environmental standards and national responses to global climate change in many countries. In most cases, however, reforms get entrenched because of elite control.

Bardhan (2004, 2007) notes that there are two broad classes of problems: informational problems (such as uncertainty about who the potential winners are ex ante) and commitment problems (the fact that the potential gainers cannot credibly commit to compensate the losers ex post). Both problems indicate that bringing about institutional change is a country-specific process that can be addressed only by applying the sovereignty principle, not through external intervention. This point suggests that elite control can be partially constrained through information and analysis and the promotion of social accountability. Transparency and information disclosure, data collection, and analysis, including analysis of the distributional impacts of institutional reforms, are key. Opening up decision-making processes to public scrutiny; ensuring the rule of law (by enforcing contracts and protecting property rights, for example); and putting anticorruption mechanisms into place are important tools for achieving social accountability. This suggests a broader approach to participation in institution-centered SEA than is typically found in traditional SEA.

Policy Making Is a Complex—and Inherently Political—Process

Political economy factors themselves often constrain transparency and accountability. As a result, change is often incremental over long periods of time. Elite capture means that directly consulting the vulnerable does not guarantee that policy will be influenced. Rules, legislation, traditions, networks, ethnic alliances, patronage, political allegiances, and bureaucratic structures interact to form a complex and fluctuating policy environment. Individual survival in an organization, organizational survival in a government, and the maintenance of a regime within a country can also affect policy choice (Grindle and Thomas 1991). Effecting change is difficult, as suggested by the conceptual framework developed by North, Wallis, and Weingast (2006), according to which the interactions between political and vested interest actors explain why limited-access social orders are a natural state. Hence, political economy analysis is an important tool for understanding the political dimension and facilitating effective policy implementation.

One way of facilitating change is to reduce regulatory capture. Regulatory capture occurs when interest groups exert undue influence on the authorities, so that instead of furthering social welfare, they further the interests of select groups.[10]

Sometimes overcoming informational problems—by, for example, providing clearer information on the distributional effects of policies—may help different stakeholders (with different positions and power) coalesce to support a policy. This may be particularly opportune for several environmental issues, because the environment is a global good. One example is urban air quality. Poor roadside vendors may suffer disproportionately from air pollution, but other more-powerful interest groups are also affected and can join with them to promote policy action. Analysis of the distributional impact of alternative policies can be important and facilitate the formation of such coalitions.

Proposed Policy SEA Approach

In order to more effectively influence policies, the proposed SEA approach aims to strengthen the institutional, technical, and governance dimensions of policy-making processes. In addition to applying analytical and participatory approaches typical within SEA frameworks, SEA for policies or institution-centered SEA also needs to incorporate a third pillar, designed to enhance learning and continuous improvement of policy design and implementation.

Analytical Work

Analytical work that is likely to facilitate the influence of SEA on policy design and implementation consists of three components: identification of environmental priorities, technical analysis, and institutional analysis.

Identifying environmental priorities. The strategic character of SEAs is established when environmental priorities are identified, determining and revealing the objectives, purposes, or goals to be pursued by public policies. The SEA approach proposed here points out the importance of periodic reevaluation of environmental priorities within a country or a sector through rigorous, quantitative analysis as well as consultation with weak and vulnerable stakeholders. This suggests the need for a priority-setting step that is both linked to the design and implementation of a specific policy and used to raise attention to priority environmental issues more broadly, so that as public policies are developed and implemented, these high-priority environmental issues are internalized in those policies. This is a major departure from traditional SEA methodologies. Several factors can influence the establishment of environmental priorities. These include public clamor; cultural, historical, institutional, and political factors; development agency priorities; international agreements; judicial decisions; and the results of technical studies employing analytical/quantitative techniques. Two quantitative techniques relevant for priority setting are cost of degradation studies and comparative risk assessments (see chapter 4). Cost of degradation studies, which use the language of finance officials, have been particularly effective in alerting senior policy makers in finance and planning ministries to the importance of environmental issues affecting the economy (Sarraf 2004; Pillai 2008).

Priority-setting exercises also focus attention in the context of scarce resources and competing priorities. As the ultimate goal to which countries aspire is long-term sustainable economic growth, it is important that any analytical work on prioritization evaluate environmental priorities, taking into account how they affect broader sustainable economic development. This suggests a departure from traditional SEA methodology, which focuses on significant stand-alone environmental impacts of a specific policy rather than its linkages with long-term sustainable economic development.

Such priority setting could be carried out by country nationals or external actors. Ideally, such capacity needs to be built through domestic universities and think tanks, but capacity building is a slow process.[11] Practice shows that the use of these tools is often driven by development agency requests for funding.[12]

Also important is the need to amplify the voice of the weak and vulnerable stakeholders in the priority-setting exercise. This can be done through assessing the distributional costs of environmental problems. Furthermore, combining quantitative methods with participatory methods allows the understanding of how environmental degradation may affect vulnerable groups to be deepened. Indeed, in traditional SEA methodology, consultation with affected stakeholders is the mechanism typically used to allow vulnerable stakeholders to voice their views. However, because of the role of vested interests and elite capture, it is important to acknowledge that conducting a one-off consultation in the context of preparing an SEA report, no matter how well done, does not ensure that reform proposals are taken up or implemented (Sánchez-Triana and Enriquez 2006). (This issue is addressed below.)

Technical analysis. Once priorities are identified, technical analysis to select effective and efficient interventions comes into play. One important aspect of this analysis is the assessment of the costs and benefits of alternative policy designs, which provides information decision makers can use if opportunities for policy reform arise. However, identifying alternative interventions and estimating the costs and benefits of each intervention involves a number of uncertainties.

As Pindyck (2007: 49) notes, environmental policy interventions "must contend with highly nonlinear cost functions, irreversibilities, and long-term horizons. But in addition, environmental problems usually involve three compounding levels of uncertainty—uncertainty over the underlying physical or ecological processes, uncertainty over the economic impacts of environmental change, and uncertainty over technological changes that might ameliorate those economic impacts and/or reduce the cost of limiting the environmental damage in the first place."

Pindyck's observation points to the importance of pairing technical analysis with institutional analysis, so that institutionally feasible interventions can be identified. Given the uncertainties in costs and benefits associated with environmental problems, the co-implementation of the pillar on learning processes (described below) helps ensure that the interventions selected steer policy design and implementation in the direction of greater environmental sustainability.

Institutional analysis. Elements that may yield better understanding of institutions linked with a particular policy reform include the following:

- Historical analysis to understand how current priorities have become locked in
- Political economy analysis, including goals, values, behaviors, and incentives of stakeholders involved in policy formulation and implementation
- Analysis of horizontal (intersectoral) and vertical (across federal, state, and municipal levels) coordination mechanisms within government to better understand implementation hurdles
- Analysis of mechanisms to promote social accountability and learning
- Identification of efficient and politically feasible interventions to overcome environmental priority issues.

Participatory Approaches

To be inclusive, participatory approaches should identify weak and vulnerable groups and amplify their voice in policy formation. Doing so will increase the likelihood that policies that are responsive to many different segments of society are designed and implemented.

Identifying weak and vulnerable stakeholders. In the context of environmental issues, vulnerable groups include those susceptible to increased health risks associated with environmental factors, those whose livelihoods are threatened as a result of changes in the natural resource base, and those susceptible to economic losses as a result of disasters ranging from flooding to toxic chemical releases to global climate change.

Health outcomes associated with environmental factors, such as respiratory diseases or waterborne diseases, are directly linked to human capital and hence current and future productivity. Most of the estimated 3 million deaths a year caused by air pollution and waterborne diseases occur in women and young children from poor families that lack access to safe water, sanitation, and modern household fuels (Ezzati and others 2004). Even in the context of global environmental issues, such as climate variability, it is the poor who often take the hardest hit, because they have limited coping strategies and access to other resources. The arguments that environment is a global good and that the environment is closely linked with improved quality of life and well-being for the poor in particular are not mutually exclusive.

The World Commission on Environment and Development's definition of sustainable development implies ensuring that future generations have at least the same opportunities as current generations. For many, the future is distant and difficult to link to present needs. It is, however, directly linked to the present through children, the future generation (Ahmed and Sánchez-Triana 2004). Children are also the ones usually most affected by environmental degradation in developing countries, principally through the impact on their health. Diarrhea, for example, causes an estimated 1.7 million deaths a year (Pruss-Ustan and others 2004).

In an institution-centered SEA, the process of identifying the most vulnerable groups differs from that used in traditional SEA methodologies, which focus on identifying groups affected by physical environmental impacts, independent of their vulnerability. The Argentina Water Supply and Sanitation SEA illustrates how the analysis provided by the SEA was targeted to design reforms aimed at providing basic urban services to the lowest income groups. It identified how lack of access to water and sanitation primarily affected poor households, with children under five facing particularly severe risks. The SEA also looked at the factors that limited expansion of coverage to poor communities (Sánchez-Triana and Enriquez 2005, 2007).

Enhancing the voice of environmentally vulnerable groups. Sustainable policies are those that favor poverty alleviation, economic growth, and control of environmental degradation. Bourguignon and others (2002, 2004), De Ferranti and others (2004), and the World Bank (2005c) all note that equity (defined primarily as equality of opportunity) is highly effective for poverty reduction, because it tends to favor sustained overall development and provides increased opportunities to the poorest groups in society.

The development literature has increasingly recognized the importance of providing an opportunity for the vulnerable to voice their needs (World Bank 2002b, 2005c). Kende-Robb and Van Wicklin argue convincingly in chapter 5 that giving the most vulnerable a voice helps policy makers understand the synergies among environmental goals, economic growth, and poverty reduction (World Bank 2000b). They describe seven tools available to amplify the voice of the most vulnerable. They also describe two case studies, in Brazil and Mongolia, where the voice of the vulnerable influenced policy formulation. They acknowledge that without political support from those in power, the voice of the vulnerable would not have been heard, much less acted upon.

In general, open political environments provide opportunities for social inclusiveness around policies for improving the quality of life and reducing poverty. In Costa Rica, where there is a tradition of bringing marginal groups into the political sphere, the government was eager to better understand poverty from the perspective of the poor and welcomed a participatory poverty assessment (Robb 2002). If a government is not fully committed to consulting with vulnerable groups, it is unlikely to act on research results that run counter to its own interest. In such circumstances, other mechanisms need to be considered for increasing the likelihood that the views of the most vulnerable are heard.

One option is the use of national advocates on behalf of vulnerable groups. In countries where the culture of consultation is weak, consulting national advocates may be easier than consulting vulnerable groups themselves, because the advocates are nationals rather than external actors. Ensuring that CSOs or other advocates, such as university professors, are present in the community of participation on behalf of vulnerable groups is one way of making their views heard. CSOs often have their own narrow agendas, however, which may not be in line with the needs of vulnerable groups. Furthermore, it is not possible to encourage CSOs to represent only vulnerable stakeholders. Encouraging CSO engagement in policy making may give powerful stakeholders another mechanism with which to influence the policy debate.

An effective CSO advocate can be found in the case study, reported by Blair in chapter 6, on urban air pollution in Delhi. In this case two fairly elitist NGOs, the Indian Council for Enviro-Legal Action and the Centre for Science and Environment, brought public interest lawsuits and waged a high-profile, fact-based publicity campaign that compelled the government to enforce legal regulations on air pollution.

Reinforcing social accountability. Social accountability refers to the obligation of public officials and decision makers to render account to citizens and society at large regarding their plans of actions, their behavior, and the results of their actions (Ackerman 2005). At a general level, accountability mechanisms include free and fair elections, legislative oversight, administrative supervision, financial audits, legal redress (the rule of law), and a free and active media.

Social accountability mechanisms refer to the broad range of actions and mechanisms (beyond voting) that citizens can use to hold the state accountable. They include citizen monitoring of public services, participatory expenditure tracking, social auditing, independent budget analysis, and civil society monitoring of the impact of policies. These initiatives regularly rely on actions by government, the media, and other societal actors that increase transparency, improve access to public information, or enhance the enabling environment for civic engagement (Malena 2005). As Blair notes in chapter 6, "If developing countries are to craft and nurture sustainable policy initiatives that can address externalities in ways that will help the environment, they will need long-term constituencies that want to support such policies and can hold policy makers accountable for their performance in implementing them. Transparency will be the critical quality in the policy process needed for these constituencies to demand accountability from policy makers."

Traditional SEA methodologies encourage disclosure of information and public participation in the context of SEAs. In contrast, the methodology proposed here emphasizes the strengthening of underlying legislation on information disclosure and public participation and related implementation practice consistent with the key messages put forth in Principle 10 of the Rio Declaration and emphasized in the 1998 Aarhus Convention on Access to Information, Public Participation in Decision-Making and Access to Justice in Environmental Matters[13]—namely, public disclosure of information, public participation in decision making, and access to justice on environmental matters. These areas are crucial for shaping a culture of inclusion and consensus building and hence ensuring social accountability for improved environmental governance.

Learning Processes

The third pillar for effective incorporation of environmental considerations into policy aims at enhancing social learning processes in order to periodically reevaluate policy direction and implementation in order to improve the quality of life of all people. In doing so, the SEA becomes more tightly focused on designing and implementing public policies that are sustainable in the long term rather than on minimizing or mitigating short-term biophysical environmental impacts.

Policy making is an iterative and incremental process. The challenge is to put in place learning mechanisms that help ensure that incremental change is directed toward long-term sustainable development. Putting learning mechanisms in place is complicated, as Ebrahim notes in chapter 7. He suggests that policy learning may be enhanced where agendas and priorities are shaped by multiple stakeholder perspectives and that politicization of an issue in broad developmental and social terms (rather than only in environmental terms) is crucial for repeated policy attention. He also suggests that policy learning may benefit from participation of stakeholders at all stages of the policy process—from agenda setting through implementation, monitoring, and evaluation. Accountability in policy learning and implementation relies not only on monitoring and evaluation but also on effective transparency and enforcement mechanisms, where transparency refers not only to providing information before key decisions are made but also to making information available in forms that are understandable to users, thus allowing for a sustained and informed policy debate and, over time, social learning.

Building on the chapter by Ebrahim, this chapter proposes that the learning pillar of institution-centered SEA facilitate the following elements of social learning:

- "Politicizing" environmental issues, by linking them to broader issues of economic and social development and poverty eradication
- Integrating agendas of environmental ministries with those of more influential ministries
- Strengthening policy advocacy networks to ensure that diverse perspectives are repeatedly placed on policy makers' agendas
- Strengthening the capacity for policy research and analysis across all types of policy actors (governments, CSOs, industry, universities, and so forth)
- Creating public forums for policy debate with appropriate backing from top-level officials to effectively create political space for policy debate
- Encouraging participation of multiple public agencies on environmental policy issues to reconcile interagency tensions while improving coordination and resources
- Putting effective transparency mechanisms in place to make information available to citizens in ways that can influence their political choices
- Monitoring and evaluating the policy-making and implementation processes to facilitate learning
- Supporting media scrutiny of policy and implementation to strengthen accountability
- Maintaining stakeholder engagement and networks after completing a policy-formulation exercise in order to ensure long-term learning and dialogue.

Learning involves generating knowledge by processing information or events and then using that knowledge to cause behavioral change. This progression is a shift from traditional SEA methodologies, which focus on predicting impacts. The proposed approach in institution-centered SEA is consistent with adaptive management (as described in chapter 3). Dialogue among actors, a culture of questioning and scrutiny, social accountability mechanisms, and constant evaluation are also important in fostering social learning.

Ultimately, promoting social learning in environmental policy appears to be less about developing technical measures or benchmarks and monitoring systems and more about creating a culture of stakeholder involvement and scrutiny among policy makers and implementers. Improving policy learning—technical, conceptual, and social—relies on enhanced communication and dialogue and continued evaluation.

Which agency should be responsible for implementing learning mechanisms? The answer is country specific. The role could be carried out by a powerful environmental agency, a planning agency, an auditor general's office, or a combination of them, among other options.

Social learning enables attention to be directed to priority environmental issues. It allows for incremental improvements over time, as states iterate through policy formulation and implementation with continually broader perspectives and increased knowledge on how to achieve more-sustainable environmental outcomes.

Lessons from Good Practice

Drawing from good practice on different components of the suggested policy SEA approach, this section illustrate the opportunities to widely apply the methodology proposed toward attaining effective designs and implementation of sustainable policies.

Setting Priorities to Periodically Reevaluate Goals and Raise Attention

In order to bring an issue to the policy agenda, it is essential to first generate attention. An example illustrates this point. The Colombia cost of degradation study showed that the estimated number of annual premature deaths from urban and indoor air pollution exceeded 7,000—more than the number in any other category, including road accidents. This fact helped attract the attention of policy makers and the mass media to the importance of tackling air pollution, an issue that had not been high on the policy agenda (Larsen 2004; World Bank 2006; Sánchez-Triana, Ahmed, and Awe 2007).

Qualitative techniques, such as participatory poverty assessments and opinion surveys, are also important. They need to be coupled with quantitative techniques, however, to ensure that perceptions are not solely driving prioritization. A random sample survey of more than 2,600 citizens from the public and private sectors, industry, academia, and civil society was used to assess environmental priorities in the context of the Colombia CEA. It revealed that priorities varied with income as well across stakeholder groups. Among lower-income groups, environmental issues linked to improvements in air quality and reduction in noise, together with reductions in risk from natural disasters, such as flooding and landslides, were most important. By contrast, higher-income groups indicated that the most important priorities were global environmental issues, biodiversity, and urban environmental issues. The general public, including many industry stakeholders, identified air pollution as the most important environmental issue, whereas environmental officials identified loss of biodiversity as most critical (CNC 2004; World Bank 2006; Sánchez-Triana, Ahmed, and Awe 2007).

Ensuring that the Voice of the Most Vulnerable Is Heard

The Government of Peru and the World Bank (2007) carried out a Country Environmental Analysis (CEA) through a participatory process to build consensus on the importance, scope, and methodologies used in the environmental analysis. The CEA process ensured that the voice of the most vulnerable was heard at several workshops attended by representatives of various sectors, including the environment, health, finances, agriculture, energy, and mining; regional environmental authorities; the private sector; NGOs; indigenous communities; civil society; and international organizations. As a result of the CEA, environmental health issues—Peru's top environmental priority—were incorporated into this analysis.[14]

As in the Colombia CEA example described earlier, distributional analysis was also used in Peru to estimate the impact of environmental degradation on the most vulnerable groups, thus helping identify linkages between environmental health issues and poverty. As elsewhere, the poor in Peru tend to be exposed to greater environmental risks than higher-income groups and to lack the resources to mitigate those risks. The analysis found that the impact of urban air pollution relative to income was more severe for the poor than for the nonpoor. Health impact relative to income was a useful indicator, because illness and premature mortality result in medical treatment costs and lost income, in addition to pain, suffering, and restriction of activity. Based on this indicator, health impacts were 75–300 percent higher among the poor than among the nonpoor (World Bank 2007). Through a media campaign, the results of the CEA analysis are helping create awareness, generate national consensus, and build broader constituencies around an issue that matters most to the vulnerable groups.

Reinforcing Social Accountability

In recent years India has made major strides in increasing access to the judiciary to address environmental pollution issues. A landmark case on air quality in Delhi firmly brought the issue to the attention of government policy makers (see chapter 6). In the early 1990s, an Indian nongovernmental organization (NGO) asked the Supreme Court to compel the Delhi government to enforce the clean air laws that had been passed some 15 years earlier. After a long and sustained campaign by the NGO, which used quantitative information on health damage effects, including estimated mortality rates, as well as an effective public awareness campaign through the press, in 1998 the Supreme Court issued its first comprehensive mandate for tackling air pollution.

The trend in India continues, with the filing of a civil writ petition linked to the installation and operation of incinerators for biomedical waste by an NGO against several government organizations, including the Ministry of the Environment and Forests and the Central, Delhi, and Maharashtra Pollution Control Boards. Environment and Forests Secretary Prodipto Ghosh (2004) has stressed the importance of voice and accountability mechanisms, such as India's strong, independent judicial institutions and free press, in harmonizing environmental concerns and economic growth.

Mechanisms to disseminate information in a manner that is easily interpretable can allow communities to function as informal regulators. Such mechanisms also promote accountability on the part of those being regulated. An example is the pioneering public disclosure scheme in Indonesia (PROPER), which encouraged firms to clean up their air and water pollution (World Bank 2000a) (see chapter 6). In a second phase of the program, the government made the disclosure program compulsory (Leitmann and Dore 2005). A mandatory program may induce greater social accountability than a voluntary program.

Other examples of accountability mechanisms include actions implemented by the government in the Mexico Programmatic Environment Structural Adjustment Loan (World Bank 2002a). These examples include public disclosure of funds returned to municipalities for water treatment investment programs to encourage greater scrutiny and accountability on the part of the public and the requirement to post on Internet the processing status of all environmental licenses. The requirement is intended to improve transparency of government procedures, thereby reducing corrupt practices. A transparency law passed in 2001 greatly facilitated these actions.

CSOs can be a key mechanism for providing voice to those who are vulnerable to environmental damage. But it is essential to fully understand the institutional context in which CSOs operate. Blackman and others (2004) note that even though Colombia's 1991 constitution and Law 99 of 1993 create numerous mechanisms for public participation in both the formulation and implementation of environmental policy, regulatory capture by private sector interest groups and political considerations have a much stronger influence. They emphasize different courses of action to strengthen the environmental NGO sector in Colombia, including promoting environmental education, making environmental data easily available, ensuring that NGOs are adequately represented on both formal and informal deliberative bodies, and adopting reforms that strengthen advance notice of significant environmental policy actions and provide opportunities for public input.

Incorporating Learning in Environmental Policy Making and Implementation

In the 1970s, the conventional wisdom was that high ambient concentrations of total suspended particles represented a serious health problem. Improvements in measurement technologies and analytical techniques have since revealed that fine particles with diameters of 2.5 microns (PM 2.5) or less appear to be the real culprits. This finding has lead to significant changes in air pollution control strategies in the United States and other countries. In Colombia, for example, through dialogue over the CEA during 2004, the importance of fine particles on health impacts was increasingly recognized, and the government has committed to move ahead with the installation of a monitoring system for PM 2.5.[15] This illustrates the importance of ensuring that systems of evaluation are able to adjust to new developments in science, technology, and other fields.

The story of indoor air pollution is still at an early stage. As recently as 2000, a special theme paper published in the *Bulletin of the World Health Organization* suggested that indoor air pollution was a major public health hazard (Bruce, Perez-Padilla, and Albalak 2000). Indoor air pollution subsequently appeared on the list of top 10 causes of illness and death in the World Health Organization's Global Burden of Disease report (WHO 2002). WHO estimates indicate that indoor smoke from solid fuels causes 1.5 million deaths a year (WHO 2007). This problem has existed for generations. Only with the creation of new information and the processing of that information did it become clear that this form of pollution affects millions of people, principally women and children in poor rural families, who depend on firewood for cooking and heating. Some countries are realizing the importance of this problem and acting on it. Many others have still not grasped the need to place this issue high up on the policy agenda.[16]

Another example of incremental behavioral change is the evolving policy in Mexico with respect to improving the sustainability of water resource management. Allocated water rights far exceed water availability in certain water-stressed areas. Water availability and quality is a crucial resource for growth and quality of life in some states. Politically, therefore, water is a contentious issue. Hence, a discussion on reducing existing "rights" to such a resource is fraught with difficulties in the short term, despite its importance from a sustainability perspective, where water is considered a potential limiting factor to future growth. Mexico's approach in addressing this issue has been incremental. The first step was to effect behavioral change by publishing water availability data and improving the water rights registry. In 2002 the possibility of "buying back" excess water rights to accurately reflect water scarcity was mere speculation (World Bank 2002a). Three years later, the government was actively buying back water rights in one overexploited aquifer (World Bank 2005b).

Other examples of behavioral change effected in a short period of time are the transformations in Bogotá, Colombia, and Curitiba, Brazil. Both cities changed expectations with respect to the quality of city life among their citizens. In Bogotá an increase in green spaces, pedestrian walkways, and bike paths; the introduction of a bus mass transit system; and the introduction of an annual "day without a car" during the short administration of Mayor Enrique Penalosa in the late 1990s changed citizens' expectations regarding the quality of city life.

Publicly available monitoring and evaluation systems are crucial, not only for technical learning but also for purposes of democratic legitimacy and public confidence. Such systems involve both ex post evaluations and ex ante assessments of policy making and impacts built on broadly shared sustainable development goals. Sometimes things do not seem as they appear. An example is the common misconception of the critical factors behind the success of Colombia's Cauca Valley Corporation's successful water pollution control program, which has led the corporation to emphasize cooperation with industry, deemphasize strict enforcement of regulations, and experiment with effluent charges. A more-thorough analysis demonstrates that an important factor for the success of the program was citizen pressure, negative publicity, and the policies of parent transnational companies calling for the use of environmental audits to facilitate compliance with environmental rules (Sánchez-Triana and Ortolano 2005).

Conclusions and Future Directions

This chapter proposes a new conceptual and methodological framework for the application of SEA to policies: institution-centered SEA. The approach extends the traditional two-pillar analytical and participatory approach in SEA methodology by the addition of a third pillar designed to enhance learning and continuous improvement of policy design and implementation. The approach takes into account the process of policy formation and the importance of seizing opportunities for policy reform as they arise. Elements within the approach also seek to enhance the creation of windows of opportunities for future policy reform.

The first pillar of the SEA approach includes three components:

- Environmental analysis, designed to identify priority environmental issues within a country or a sector through quantitative analysis and consultation with weak and vulnerable stakeholders
- Technical analysis, which includes assessment of the costs and benefits of alternative policy designs
- Institutional analysis, which includes historical analysis to understand how current priorities have become locked in; distributional costs of environmental problems; political economy analysis, including goals, values, behaviors, and incentives of stakeholders involved in policy formulation and implementation; analysis of mechanisms to promote social accountability and learning; analysis of horizontal (intersectoral) and vertical (across federal, state and municipal levels) coordination mechanisms within governments to better understand implementation hurdles; and identification of efficient and politically feasible interventions to overcome environmental priority issues.

The second pillar seeks to enhance social accountability of public officials to all citizens, including the weak and vulnerable, and to facilitate the creation of windows of opportunity for continuous improvement of policies. The approach, therefore, suggests that institution-centered SEA move from consultation of affected groups to amplifying the voice of vulnerable groups through institutional mechanisms, in order to facilitate the design of policies that are responsive to many different segments of society.

The third pillar aims to enhance social learning processes so that there is periodic reevaluation of policy direction and implementation in order to improve the quality of life of all citizens. The focus is on designing and implementing public policies that are sustainable in the long term rather than on minimizing or mitigating short-term biophysical environmental impacts.

The outputs from institution-centered SEAs can be divided into two groups. The first are those with short-term impact. Identifying these recommendations is particularly useful if decision makers are to be able to seize opportunities as windows for reform open up. Recommendations with longer-term impact help strengthen institutional mechanisms over a longer period, so that public officials are socially accountable to their citizens and can continue to pursue continuous improvement of public policies in a direction of greater sustainability. This approach, in turn, helps create windows of opportunity for future reform efforts.

Several countries have established national legislation and regulations that require that SEAs be carried out by governmental agencies. SEA legal instruments include the European SEA Directive and the SEA Protocol to the UNECE (Espoo) Convention on Environmental Impact Assessment in a Transboundary Context. These national and regional legal instruments focus on SEA for plans and programs. Several countries, including Canada, the Dominican Republic, Kenya, the Netherlands, and South Africa, also require SEA for policies (Ahmed and Fiadjoe 2006). Current SEA requirements in most national legal instruments emphasize traditional impact-centered SEA methodologies (Ahmed and Fiadjoe 2006). If the proposed approach to SEA for policies presented in this chapter is validated, countries developing legislation requiring application of SEA to policies may wish to rethink how SEA is described in their national legal instruments.

There is also increasing push for the application of SEA, by both donors and developing countries, to make aid to developing countries more effective (OECD 2006). Given scarce human and financial resources in developing countries, external resources from development aid could be harnessed to support developing countries' efforts to conduct such assessments. Some examples include refining the broader country-level environmental analyses carried out by development agencies (which already include some level of assessment of both priority environmental problems and institutional aspects) to better incorporate these concepts. Assessments of transparency, participation, and access conducted by CSOs could also be drawn on.[17]

If a country is to design and implement policies that enhance synergies among poverty alleviation, economic growth, and a healthy environment that enhance the quality of life, it must be able to learn from experience. An institution-centered SEA is likely to be more effective in influencing policy in a country with a strong learning framework and culture. However, in countries with weaker learning frameworks, an institution-centered SEA that provides information and nurtures internal champions and the formation of constituencies to enhance learning could also play an influential role.

Two broad aspects are particularly important for enhancing learning. In line with adaptive management are monitoring and evaluation frameworks, which continuously reexamine the priorities and underlying goals, evaluate distributional impacts of environmental problems, assess distributional impacts of policy implementation, and bring to the attention of policy makers the results of monitoring and evaluation. In line with inclusive management are promoting a culture of debate, scrutiny, and social accountability through information dissemination, using different mechanisms and forums to bring stakeholders together to participate in decision making, and access to recourse mechanisms or justice on environmental matters.

Both require long periods of continuous engagement by policy makers as they make changes in formal rules and attempt to change informal rules (culture, behavior, and so forth). For this reason, the effectiveness of institution-centered SEA can be evaluated only over a long period of time. Such evaluation should assess both immediate changes in policies that improve environmental sustainability or quality of life, as well as the ability of the SEA to support decision makers in strengthening learning frameworks.

The World Bank is currently supporting a pilot program to test this approach in various countries and sectors. Evaluation of this program in the coming years will yield important lessons for refinement of this approach. However, if validated, the most important implication is that institution-centered SEA could help developing country governments move toward a path that integrates poverty alleviation, social equity, economic growth, and ultimately environmental sustainability.

Notes

1 The authors, who task managed the Colombia Country Environmental Analysis, view this product as an example of an institution-centered SEA.

2 The CARs were set up in Colombia between 1952 and 1974 as autonomous agencies with responsibilities for implementing water resources and infrastructure projects (based on the model of the Tennessee Valley Authority in the United States) (Sánchez-Triana 2007). By 1980, their mandate was expanded beyond these economic development functions to include environmental regulation (Sánchez-Triana and Ebrahim 1999; Sánchez-Triana, Ahmed, and Awe 2007).

3 The six impact indicators for environmental quality in the PATs cover issues such as deforestation rates and forest conservation efforts, development of green markets, rationalization and optimization of renewable natural resource consumption, reduction in health impacts associated with environmental factors, and reduction in vulnerability risk associated with natural disasters.

4 See http://www.worldbank.org/seatoolkit for more details.

5 This assessment was conducted in 2005, based on information available at the time. It does not take into account any changes to these SEA systems made since June 2005.

6 The developmental path of policies is usually conditioned by those policies' history. David (1985) and Arthur (1988a, 1988b, 1989, 1994) used the technologies they studied to

demonstrate "the peculiar fact that incremental changes in technology, once begun on a particular track, may lead one technological solution to win out over another, even when, ultimately this technological path may be less efficient than the abandoned alternative would have been" (North 1990: 93). North (1990) extends Arthur's concept of path dependence to institutional change and organizational decision making. For him (1990) path dependence occurs because "the present and the future are connected to the past by the continuity of society's institutions. Today's and tomorrow's choices are shaped by the past" (North, 1990: vii).

7   Easterly (2006) alludes to this in his depiction of the effectiveness of "planners" versus "preachers" in the context of development aid.

8   Fearing replacement, the monarchy and aristocracy in 19th century Russia and Austria-Hungary blocked the establishment of institutions to facilitate industrialization (Acemoglu and Robinson 2006). The divergent development paths in North and South America since colonial times suggest how, in societies with high levels of inequality before colonization, institutions evolved in ways that restricted access to political power and economic opportunities to a narrow elite (Engerman and Sokoloff 2002). Initial unequal conditions had long-term effects through elite influence on public policies.

9   A recent case study of the Bogotá Savanna documents that powerful interests, such as the irrigation district, the regional environmental corporation, and the energy company, have secured access to a sufficient quantity of water in the Savanna. In contrast, small farmers in the region lack adequate water supplies (World Bank 2005a).

10   The National Technical Advisory Council in Colombia was set up under Article 11 of Law 99 of 1993 to assist the Ministry of the Environment in assessing the technical feasibility of environmental projects, policies, and regulations. It is directed by a secretary appointed by the minister. Its members include representatives from universities, as well as representatives from agriculture, mining, the petroleum industry, and other industries. This council is an example of regulatory capture, because private sector interests dominate (Blackman and others 2004).

11   The World Bank assisted the Government of Mexico in building national capacity for environmental economic analysis by creating and strengthening the Economic and Social Unit in the Mexican Ministry of the Environment and Natural Resources under the National Environmental Project, initiated in 1992 (World Bank 2005a). Sarraf (2004) reports that the Tunisian National Environmental Protection Agency is interested in setting up a unit of economists who will conduct economic analyses of environmental projects.

12   It was originally envisioned that client countries would finance and implement Poverty and Social Impact Analysis (PSIA) (personal communication Anis Dani, Advisor, Social Development Department, World Bank, May 17, 2005). The more than 100 PSIA studies initiated between 2001 and 2005 revealed the human and financial resource limitations to their doing so, however. Although governments have taken the lead in designing and using the results of these studies, capacity constraints and resource limitations often prevent them from managing them.

13.   See http://www.unece.org/env/pp/documents/cep43e.pdf.

14   Air pollution and other environmental health problems were traditionally not considered priority issues in the environmental agenda in Peru.

15   Another air pollution issue that illustrates behavioral change is the phasing out of lead in gasoline as a result of greater understanding of its health impacts on children, as well as better knowledge that is based on other country's experiences about how to move forward and implement such a policy.

16 In Guatemala information on the connection between indoor pollution and health was lacking at all levels, from senior policy makers to poor women, many of whom were unaware of the link between health and smoke from indoor combustion of fuelwood (Ahmed, Awe and others 2005).

17 See, for example, http://www.accessinitiative.org or http://www.pp10.org.

References

Acemoglu, Daron, and James A. Robinson. 2006. "Persistence of Power, Elites and Institutions." NBER Working Paper Series W12108, National Bureau of Economic Research, Cambridge, MA.

Ackerman, John M. 2005. *Social Accountability in the Public Sector: A Conceptual Discussion.* Social Development Paper 82, World Bank, Washington, DC.

Ahmed, Kulsum, Yewande Awe, Douglas Barnes, Maureen Cropper, and Masami Kojima. 2005. *Environmental Health and Traditional Fuel Use in Guatemala.* Directions in Development Series, Washington, DC: World Bank.

Ahmed, Kulsum, and Yvonne Fiadjoe. 2006. "A Selective Review of SEA Legislation: Results from a Nine Country Review." Environment Strategy Paper 13, World Bank, Washington, DC.

Ahmed, Kulsum, Jean-Roger Mercier, and Rob Verheem. 2005. *Strategic Environment Assessment: Concept and Practice.* Environment Strategy Note 14, World Bank, Washington, DC.

Ahmed, Kulsum, and Ernesto Sánchez-Triana. 2004. "Sustainable Development and Policy Reform: Implementing MDG 7, Target 9." *Environment Matters*, World Bank Annual Review (July 2003–June 2004), 22–25.

Arthur, W. Brian. 1988a. "Competing Technologies: An Overview." In *Technology and Economics*, eds. G. Dosi, C. Freeman, R. Nelson, L. Soete, and G. Silverberg. London: Pinter.

———. 1988b. "Self-Reinforcing Mechanisms in Economics." In *The Economy as an Evolving Complex System*, eds. K. J. Arrow and P. Anderson, 9–33. New York: John Wiley.

———. 1989. "Competing Technologies, Increasing Returns, and Lock-In by Historical Events." *Economic Journal* 99 (394): 116–31.

———. 1994. *Increasing Returns and Path Dependence in the Economy.* Ann Arbor: University of Michigan Press.

Bardhan, Pranab. 1989. "The New Institutional Economics and Development Theory." *World Development* 17 (9): 1389–95.

———. 2004. *Scarcity, Conflicts, and Cooperation Essays in the Political and Institutional Economics of Development.* Cambridge, MA: MIT Press.

———. 2007. "Institutional Economics of Development: Some Reflections on Recent Advances." DEC Lecture, April 19, World Bank, Washington, DC.

Blackman, Allen, Sandra Hoffman, Richard Morgenstern, and Elizabeth Topping. 2004. *Assessment of Colombia's National Environmental System (SINA).* Washington, DC: Resources for the Future.

Bourguignon, François, Francisco Ferreira, and Nora Lustig. 2004. *The Microeconomics of Income Distribution Dynamics in East Asia and Latin America.* New York: Oxford University Press for the World Bank.

Bourguignon François, and Christian Morrisson. 2002. "Inequality among World Citizens: 1820–1992." *American Economic Review* 92 (4): 727–44.

Bruce, Nigel, Rogelio Perez-Padilla, and Rachel Albalak. 2000. "Indoor Air Pollution in Developing Countries: A Major Environmental and Public Health Challenge." *Bulletin of the World Health Organization* 78 (9): 1078–92.

Chavarro Vásquez, Pedro A. 2007. *Evaluación del impacto del Análisis Ambiental de País (AAP) en las políticas gubernamentales y percepción públicas.* Report commissioned by the World Bank, Latin America and the Caribbean Region, Washington, DC.

CNC (Centro Nacional de Consultoria). 2004. "Colombian Environmental Priorities." Presentation made at the Country Environmental Analysis Workshop, Bogotá, August 26.

Dalal-Clayton, B., and B. Sadler. 2005. "Strategic Environmental Assessment (SEA): A Sourcebook and Reference Guide to International Experience." London: Earthscan.

David , Paul A. 1985. "Clio and the Economics of QWERTY." *American Economic Review* 75 (2): 332–37.

De Ferranti, David, Guillermo E. Perry, Francisco H. G. Ferreira, and Michael Walton. 2004. *Inequality in Latin America: Breaking with History?* Washington, DC: World Bank.

Easterly, William. 2001. *The Elusive Quest for Growth.* Cambridge, MA: MIT Press.

————. 2006. *The White Man's Burden.* London: Penguin Press.

Engerman, Stanley L., and Kenneth L. Sokoloff. 2002. *Factor Endowments, Inequality, and Paths of Development among New World Economies.* NBER Working Paper Series 9259, National Bureau of Economic Research, Cambridge, MA.

Ezzati, Majid, Alan D. Lopez, Anthony Rodgers, and Christopher J. L. Murray. 2004. *Comparative Quantification of Health Risks: Global and Regional Burden of Disease Attributable to Selected Major Risk Factors,* vol. 1. Geneva: World Health Organization.

Fischer, Thomas B. 2002. *Strategic Environmental Assessment in Transport and Land Use Planning.* London: Earthscan.

Ghosh, Prodipto. 2004. "Harmonizing Environmental Concerns and Economic Growth: the Indian Perspective." *Environment Matters,* World Bank Annual Review (July 2003–June 2004): 10–11.

Grindle, M. S., and J. W. Thomas. 1991. *Public Choices and Policy Change: The Political Economy of Reform in Developing Countries.* Baltimore, MD: Johns Hopkins University Press.

Larsen, Bjorn. 2004. *Colombia: Cost of Environmental Damage. A Socio-Economic and Environmental Health Risk Assessment.* Consultant report prepared for the Ministry of Environment, Housing and Land Development, Bogotá.

Leitmann, J., and G. Dore. 2005. "Better Environmental Governance: Improving the Role of Local Governments and the Private Sector in Indonesia." Presentation prepared for the Environmentally and Socially Sustainable Development Week, World Bank, Washington, DC.

Malena, Carmen, with Reiner Forster and Janmejay Singh. 2005. *Social Accountability: An Introduction to the Concept and Emerging Practice.* Social Development Paper 76, World Bank, Washington, DC.

Noble, Bram F. 2002. "Strategic Environmental Assessment: What Is It and What Makes It Strategic?" *Journal of Environmental Assessment Policy and Management* 2 (2): 203–24.

North, Douglass C. 1990. *Institutions, Institutional Change, and Economic Performance.* New York: Cambridge University Press.

————. 1993. "Nobel Prize Lecture." December 9, Stockholm, Sweden.

North, Douglass C., John Joseph Wallis, and Barry R. Weingast. 2006. *A Conceptual Framework for Interpreting Recorded Human History*. NBER Working Paper Series 12795, National Bureau of Economic Research, Cambridge, MA.

OECD (Organisation for Economic Co-operation and Development). 2006. *Applying Strategic Environmental Assessment (SEA): Good Practice Guidance for Development Co-operation*. Development Assistance Committee Guidelines and Reference Series. Paris: OECD Publishing.

Partidário, Maria Rosario. 2002. "Strategic Environmental Assessment (SEA): Key Elements and Practices in European Approaches. Course Manual." Training course presented at the 22nd Annual Meeting of the International Association for Impact Assessment, Netherlands Congress Center, the Hague, June 15–16.

Pindyck, Robert S. 2007. "Uncertainty in Environmental Economics." *Review of Environmental Economics and Policy* 1 (1): 45–65.

Pillai, Poonam. 2008. "Strengthening Policy Dialogue on Environment: Learning from Five Years of Country Environment Analysis." Environment Department Paper Series, World Bank, Washington, DC.

Pruss-Ustan, A., D. Kay, L. Fewtrell, and J. Bartram. 2004. "Unsafe Water, Sanitation and Hygiene." In *Comparative Quantification of Health Risks: Global and Regional Burden of Disease Attributable to Selected Major Risk Factors*, vol. 1, eds. Majid Ezzati, Alan D. Lopez, Anthony Rodgers, and Christopher J. L. Murray. Geneva: World Health Organization.

Robb, Caroline M. 2002. *Can The Poor Influence Policy? Participatory Poverty Assessments in the Developing World*. Washington, DC: World Bank and International Monetary Fund.

Sadler, Barry, and Rob Verheem.1996. *Strategic Environmental Assessment: Status, Challenges and Future Directions*. The Hague: Ministry of Housing, Spatial Planning and the Environment.

Sánchez-Triana, Ernesto. 2007. "A Unique Model for Environmental Management." In *Environmental Priorities and Poverty Reduction: A Country Environmental Analysis for Colombia*. Directions in Development Series. eds. E. Sánchez Triana, Kulsum Ahmed, and Yewande Awe. Washington, DC: World Bank.

Sánchez-Triana, Ernesto, Kulsum Ahmed, and Yewande Awe, eds. 2007. *Environmental Priorities and Poverty Reduction: A Country Environmental Analysis for Colombia*. Directions in Development Series. Washington, DC: World Bank.

Sánchez-Triana, Ernesto, and A. Ebrahim. 1999. "From the Tennessee Valley to the Cauca Valley: Differential Isomorphism in the Capture of Environmental Regulatory Agencies." Working Document, Instituto de Estudios Ambientales de la Universidad Nacional de Colombia, Bogota.

Sánchez-Triana, Ernesto, and S. Enriquez. 2005. "Using Strategic Assessments for Environmental Mainstreaming in the Water and Sanitation Sector: The Cases of Argentina and Colombia." Sustainable Development Working Paper 26, Latin America and Caribbean Region, World Bank, Washington, DC.

———. 2006. "Using Strategic Environmental Assessments for Environmental Mainstreaming in the Water and Sanitation Sector: The Cases of Argentina and Colombia." Environment Strategy Note 15, World Bank, Washington, DC.

———. 2007. "Using Policy-Based Strategic Environmental Assessments in Water Supply and Sanitation Sector Reforms: The Cases of Argentina and Colombia." *Impact Assessment and Project Appraisal* 25 (3): 175–88.

Sánchez-Triana, Ernesto, and Leonard Ortolano. 2005. "Influence of Organizational Learning on Water Pollution Control in Colombia's Cauca Valley." *International Journal of Water Resources Development* 21 (3): 493–508.

Sarraf, Maria. 2004. "Assessing the Costs of Environmental Degradation in the Middle East and North Africa Region." Environment Strategy Note 9, World Bank, Washington, DC.

Uribe, Alvaro. 2002a. "Manifiesto democrático: Los 100 puntos de Uribe." *El Colombiano*, May 30, Medellin.

———. 2002b. "Punto 64. La Colombia que quiero. Manifiesto democrático: Los 100 puntos de Uribe." *El Colombiano*, May 3, Medellin.

WHO (World Health Organization). 2002. *World Health Report*. Geneva.

———. 2007. *Indoor Air Pollution: National Burden of Disease Estimates (Revised)*. Geneva.

World Bank. 2000a. *Greening Industry*. Washington, DC: World Bank.

———. 2000b. *World Bank Development Report 2000/2001: Attacking Poverty*. Washington, DC: World Bank.

———. 2002a. "Mexico: Programmatic Environment Structural Adjustment Loan: Program Document. Report 24458 ME." July 8, Washington, DC.

———. 2002b. *World Development Report 2003: Sustainable Development in a Dynamic World*. Washington, DC: World Bank.

———. 2005a. *Integrating Environmental Considerations in Policy Formulation: Lessons from Policy-Based SEA Experience*. Report 32783, Washington, DC.

———. 2005b. "Mexico: Second Programmatic Environment Development Policy Loan: Program Document." Report 32248–MX, August 9, World Bank, Washington, DC.

———. 2005c. *World Development Report 2006: Equity and Development*. Washington, DC: World Bank.

———. 2006. "Republic of Colombia: Mitigating Environmental Degradation to Foster Growth and Reduce Inequality." Report 36345–CO, Latin America and the Caribbean Region, Environmentally and Socially Sustainable Development Department, World Bank, Washington, DC.

———. 2007. *Environmental Sustainability: A Key to Poverty Reduction in Peru. Country Environmental Analysis*. Washington, DC: Latin American and the Caribbean Region, Environmentally and Socially Sustainable Development Department.

INDEX

Boxes, figures, notes, and tables are indicated by b, f, n, and t, respectively.

ECO-AUDIT
Environmental Benefits Statement

The World Bank is committed to preserving endangered forests and natural resources. The Office of the Publisher has chosen to print *Strategic Environmental Assessment for Policies* on recycled paper with 30 percent postconsumer fiber in accordance with the recommended standards for paper usage set by the Green Press Initiative, a nonprofit program supporting publishers in using fiber that is not sourced from endangered forests. For more information, visit www.greenpressinitiative.org.

Saved:
- 6 trees
- 4 million BTUs of total energy
- 500 lbs. CO_2 equivalent of greenhouse gases
- 2,076 gallons of waste water
- 267 pounds of solid waste